思科系列丛书之 Packet Tracer 经典案例篇

Packet Tracer 经典案例之路由交换进阶篇

刘彩凤　编著

电子工业出版社
Publishing House of Electronics Industry
北京·BEIJING

内 容 简 介

本书基于 Cisco 最新版本的 Packet Tracer 模拟器开发了近 80 个典型、实用的教学案例，以启发读者思考，激发读者学习兴趣，让读者深入学习。本书特色是：案例丰富，旨在增强趣味性和挑战性；拓扑精简，重在提高技术活学活用水平。全书共 8 章，主要内容包括学习交换冗余技术、学习 LAN 安全技术、学习基础 OSPFv2、学习高级 OSPFv2、学习 ACL 技术、学习 NAT 技术、学习基础 IPv6 和学习高级 IPv6。

本书既可作为思科网络技术学院的实验教材，也可作为电子和计算机等专业的网络集成类课程的教材或实验指导书，还可作为计算机网络技能大赛的实训教材，同时也是一本网络工程师和网络规划师在工作和学习中不可多得的参考书。

未经许可，不得以任何方式复制或抄袭本书之部分或全部内容。
版权所有，侵权必究。

图书在版编目（CIP）数据

Packet Tracer 经典案例之路由交换进阶篇 / 刘彩凤编著. —北京：电子工业出版社，2021.10
（思科系列丛书. Packet Tracer 经典案例篇）
ISBN 978-7-121-42162-4

Ⅰ. ①P… Ⅱ. ①刘… Ⅲ. ①计算机网络-网络设备-教学软件-高等学校-教材 Ⅳ. ①TP393

中国版本图书馆 CIP 数据核字（2021）第 204007 号

责任编辑：满美希　　文字编辑：宋　梅
印　　刷：三河市双峰印刷装订有限公司
装　　订：三河市双峰印刷装订有限公司
出版发行：电子工业出版社
　　　　　北京市海淀区万寿路 173 信箱　邮编　100036
开　　本：787×980　1/16　印张：19　字数：438 千字
版　　次：2021 年 10 月第 1 版
印　　次：2022 年 8 月第 2 次印刷
定　　价：89.00 元

凡所购买电子工业出版社图书有缺损问题，请向购买书店调换。若书店售缺，请与本社发行部联系，联系及邮购电话：（010）88254888，88258888。
质量投诉请发邮件至 zlts@phei.com.cn，盗版侵权举报请发邮件至 dbqq@phei.com.cn。
本书咨询联系方式：mariams@phei.com.cn。

序　言

　　思科公司为其网络技术学院开发的 Packet Tracer 是业界最受欢迎的模拟软件之一，该模拟软件作为网络教学和学习的辅助工具，在思科网络技术学院的教学实践中被广为使用。目前，网络技术发展日新月异，新设备、新功能层出不穷，作为一款广受欢迎的模拟软件，Packet Tracer 的版本也在不断地更新。

　　为了向用户说明新增特性与功能的使用方法，每一次新发布的软件包中都会包含一些示例文件，以帮助用户了解相应的配置方法与应用实例。这些示例文件主要来自软件开发团队，重在阐明设备功能与配置方法，示例的拓扑都比较简单。而在实际网络应用环境中，往往要综合利用各项功能，并且网络设备和拓扑多种多样。因此，用户在了解了设备的基本功能后，还需要具备综合利用各项功能构建复杂网络拓扑的能力。对于网络专业的师生来说，尤其需要一些利用 Packet Tracer 工具搭建的集成各项功能的综合应用案例，并将它们融入教学实践中。这正是 Packet Tracer 软件目前所欠缺的部分。

　　烟台职业学院的刘彩凤老师常年活跃在教学一线，并且成为思科网络技术学院的专职讲师已有 15 年的时间。刘老师多年来对 Packet Tracer 工具进行了不懈的探索与钻研，并积极将其应用于教学实践中。同时她多次受邀在思科网络技术学院年会上分享教学经验和教学案例，获得了同行的好评与认可。刘老师根据自己多年来对 Packet Tracer 软件的使用经验和教学实践，先后出版了《Packet Tracer 经典案例之路由交换入门篇》和《Packet Tracer 经典案例之路由交换综合篇》，这两册图书属于"思科系列丛书之 Packet Tracer 经典案例篇"，书中包含丰富的教学案例，每个案例都经过精心设计，并在教学实践中进行优化，正好弥补了 Packet Tracer 软件缺乏综合案例的缺陷。这两册书自出版发行以来，被各类院校的网络专业师生广泛采用，深受读者欢迎。

　　本书作为"思科系列丛书之 Packet Tracer 经典案例篇"的第三册，不但继承了前作的特点，而且其内容在前作的基础上有所拓展，案例涉及的范围更加广泛，内容更加深入。本书的案例涵盖了交换冗余技术、局域网安全、OSPFv2 高级特性、访问控制列表技术、NAT 和 IPv6 的基础和高级特性等内容，其主题紧扣网络专业课程内容，特别是 IPv6 方面的内容，配合目前国家积极推进 IPv6 部署的要求，弥补了学校教材中缺乏 IPv6 相关案例的不足。书中的案例设计遵循学习认知规律，循序渐进、由浅入深。同时案例形式多样，从案例的来源划分，既有来源于课堂教学的实验案例，也有来源于实际网络应用的应用案例，如企业园区网案例，还有直接来源于实际的工程项目案例，使读者有机会了解如何利用 Packet Tracer 工具完成复杂的大型网络设计。从案例的类型划分，既有配置类、故障排除类和探索发现类的案例，也有师生互动类、锻炼逻辑推理能力的实验验证类的案例，还有挑战闯关类的案例。读者可选择不同类型的案例进行实验实训，这些案例也可作为网络技能比赛的训练项目。书中的案例形式新颖活泼，寓教于乐，可启发读者思考，激发读者学习兴趣。这些设计独特的案例无一不是作者多年教学实践的结晶。

本书的出版是继刘老师前两部关于 Packet Tracer 的著作之后的又一项重要成果。我作为思科网络技术学院项目的见证者、参与者和推动者，特别向读者推荐本书。希望思科网络技术学院的广大教师和学生能够从"思科系列丛书之 Packet Tracer 经典案例篇"中充分汲取营养，希望本套丛书为培养更多、更优秀的网络人才发挥最大效益，同时也衷心希望思科网络技术学院项目继续推动教育事业的发展。

最后，借此机会感谢刘彩凤老师在百忙中抽出时间为读者奉献这套优秀的图书，也感谢电子工业出版社的宋梅编审为出版这套图书所付出的辛勤劳动！

李涤非

思科公司大中华区思科网络技术学院技术经理

2021 年 4 月 3 日

前 言

2021 年是我作为思科网络技术学院专职讲师的第 15 个年头，呈现给读者的这三本书是我十多年来教学实践的总结与积淀。2017 年《Packet Tracer 经典案例之路由交换入门篇》出版，2020 年《Packet Tracer 经典案例之路由交换综合篇》出版，出版几年来，广泛被各类院校使用。应读者的强烈要求，再出一部进阶篇，感恩读者的厚爱。由于教学工作繁重，家庭事务缠绕，再加上 Packet Tracer 版本不断升级，给本书的按时完成带来了一定困难。在我和我的同事及学生团队的共同努力下，2020 年暑假完成本书初稿；2020 年秋季，经过一轮教学实践；2021 年寒假我对书稿进行了修改、完善，终于使其定稿，完成了历时近两年的编写工作。

本书目标

本书基于 Cisco Packet Tracer 开发了近 80 个教学案例，其目标是帮助读者深入学习网络技术并将其灵活地应用到生活和工作中，使读者成为一名合格的网络工程师，进而打造"互联网+"时代的网络技术精英。

内容组织

案例设计遵循认知规律，由简至繁，由易至难，从交换到路由，从 IPv4 到 IPv6，从简单需求到复杂需求，从常规配置到拓展配置。全书共 8 章，各章简要内容如下。

第 1 章　学习交换冗余技术：本章主要介绍 STP、链路聚合和网关冗余等内容。

第 2 章　学习 LAN 安全技术：本章主要介绍端口安全、IEEE 802.1x 认证、DHCP 侦听、防止 ARP 攻击和 STP 优化等内容。

第 3 章　学习基础 OSPFv2：本章主要内容包括 OSPFv2 概述、基础 OSPFv2 配置、OSPFv2 验证和 OSPFv2 的灵活应用——干预 OSPFv2 运行等。

第 4 章　学习高级 OSPFv2：本章主要介绍 OSPFv2 虚链路配置、OSPFv2 认证配置、OSPFv2 路由注入配置、OSPFv2 路由汇总配置和 OSPFv2 特殊区域配置等内容。

第 5 章　学习 ACL 技术：本章主要内容包括 ACL 概述、标准 ACL 配置、扩展 ACL 配置和 ACL 应用拓展等。

第 6 章　学习 NAT 技术：本章主要内容包括认识 NAT 技术、配置多种类型的 NAT 和配置拓展 NAT 等。

第 7 章　学习基础 IPv6：本章主要内容包括学习 IPv6 基础知识、配置 IPv6 地址、配置 SLAAC 与 DHCPv6 等。

第 8 章　学习高级 IPv6：本章主要介绍 IPv6 静态路由、IPv6 动态路由、IPv6 ACL 和 IPv6 综合案例配置等内容。

本书特色

本书内容丰富，包括近 80 个富有趣味性和挑战性的案例。案例设计力求创新，设计思路循序渐进，环环相扣。案例形式新颖活泼且不失严谨务实，内容简洁清晰且不失深刻厚重。让读者在仿真环境中潜心学习，启发思考，激发兴趣。

案例独特，类型多样

案例设计，源于实践，不拘一格。有来自课堂、实验室的，也有来自企业、园区网的；有配置类的，也有排错类的；有探索发现类的，也有逻辑推理类的；有师生互动类的，也有自主完成类的；有实验验证类的，也有技能强化类的；有课内学习类的，也有挑战闯关类的。

多方选材，布局合理

案例取材，集精品项目于一体。有学生设计的精品案例，也有教师教学设计获奖作品改编的教学案例；有单一拓扑承载多技术的应用场景案例，也有拓扑变化多端的应用案例；有来自课堂的实验案例，也有比赛的训练案例；有 IPv4 案例，也有 IPv6 案例；有对可靠性要求较高的案例，也有对安全性要求较高的案例。

贴近实际，易教乐学

表现形式，融网络技术于生活。案例设计遵循认知规律，由浅入深，由简至繁，将枯燥知识生活化和故事化，借助趣味挑战案例，让读者能够深入学习理论知识，开动脑筋，熟练掌握所学技术，做到活学活用，顺利闯关。

读者对象

本书既可作为思科网络技术学院的实验教材，也可作为电子和计算机等专业的网络集成类课程的教材或实验指导书，还可作为计算机网络技能大赛的实训教材，同时也是一本网络工程师和网络规划师在工作和学习中不可多得的参考书。

阅读建议

因为各项目案例相对独立，所以建议读者在阅读本书时先参考目录，从自己感兴趣的项目入手。

特别声明

在本书项目拓扑中，如果设备间连线是虚线，表示采用交叉线；如果设备间连线是光纤，则在图中有特别标注。

在本书中，接口的名字均采用简写，其中 Gi 的全称是 GigabitEthernet，Fa 的全称为 FastEthernet，Se 的全称为 Serial，本书后续内容均采用简写来描述网络设备的接口。

Packet Tracer 是思科网络技术学院的教学工具。思科网络技术学院的教师、学生及校友都可以使用该工具辅助学习 IT 基础、CCNA 路由和交换、CCNA 安全、物联网、无线网络等课程。读者可以通过本书配套教学资源课件中提供的 Cisco Packet Tracer 的官方下载链接注册成为"Packet Tracer 101"课程的学生并下载最新版 Packet Tracer 软件。

本书配套有教学资源课件，如有需要，请登录电子工业出版社华信教育资源网（www.hxedu.com.cn），注册后免费下载。

致谢

本书由刘彩凤编写并统稿，参加编写工作的还有韩茂玲、崔玉礼、张津铭、于洋和王笑娟。感谢思科网络技术学院原全球技术总监 John Lim 及其团队，使我有机会参加全球教师资源设计竞赛（GIR Contest），吸收国内外先进教学理念，提升教学设计能力；感谢思科大中华区公共事务部总监练沛强先生，本书编写得到了练总大力支持，本书配套教学资源课件中有关 Cisco Packet Tracer 的官方下载链接由练总提供；感谢思科网络技术学院原全球产品经理刘亢，他给予我机会参与 Cisco Packet Tracer 测试，鼓励我参加基于 Packet Tracer 的教学案例设计竞赛，让我不断提高；感谢思科公司大中华区原企业社会责任经理韩江先生，让我有幸参与思科校企案例项目开发，积累素材，坚定我的创作信念，本书正是因为他的提议才诞生；感谢电子工业出版社有限公司宋梅编审，没有宋老师的鞭策和鼓励，本书与读者见面将会遥遥无期，也正是宋老师加班工作，才加快了本书的出版进度；感谢思科网络技术学院大中华区技术经理李涤非老师多年来对我的专业指导和经验传授，让我在写作上少走很多弯路；感谢思科公司中国区原公共事务部企业社会责任经理徐如滢女士，让我有机会与思科公司合作院校进行交流，一路携手，使我不断提高；感谢思科公司企业社会责任经理熊露颖先生，让我有机会参与思科英文教材翻译校验及 PT 考试系统开发，为本书编写奠定了基础；感谢思科公司原企业社会责任经理张冉先生对本书编写提出的宝贵意见。

感谢烟台职业学院院长温金祥教授对本书编写给予的支持与关注；感谢烟台职业学院副院长房培玉教授引导我走进思科网络技术学院，开启我的网络教学生涯；感谢烟台职业学院王作鹏副院长在本书创作过程中给予我的支持和指导；感谢烟台职业学院教务处长原宪瑞教授在教育理念和整体架构上给予我的指导和影响；感谢国家精品资源共享课程负责人薛元昕教授在课程建设及资源开发方面给予我的帮助；感谢烟台职业学院徐言超老师，广州黄埔职业技术学校何力老师，河南信息工程学校谢晓广老师、门亚范老师，吉林铁道职业技术学院王爱华老师，中国石油大学肖军弼老师、曹绍华老师等对本书编写工作的大力支持。

感谢我强大的学生团队［李雪林（17NET2）、于飞凡（16NET1）、马恺（18NET2）、满东豪（18NET2）、甄金强（16NET1）、王宝宇（18NET1）、胡颖（15NET2）、王军（17NET1）、王雪蕾（15NET2）、尹翠红（11NET1）、卜云霞（11NET1）、黎振（08NET2）、柳涛（05NET

等]对本书编写提出的宝贵意见和对相关技术细节的反复验证，他们对本书贡献巨大。尤其感谢我的学生李雪林、于飞凡、马恺、满东豪、甄金强，是他们陪伴我完成了本书创作。最后，感谢思科公司和思科网络技术学院，以及对本书寄予厚望的老师、历届的学生们，是他们给了我无限动力。

感言

本书创作过程非常艰辛，写作周期长，设计的案例要在实践中不断验证。为潜心创作，2020年暑假我们进行了一个多月的封闭写作。在封闭期间，通信工具时常会中断，感谢家人、朋友、同事对我的支持、理解和包容。尽管创作艰辛，但我很享受设计灵感一次次迸发的过程，期盼与大家一起分享这份成果。虽然尽了最大努力，但因本人水平和视野有限，书中难免存在纰漏和不足之处，愿读者朋友们给予指正，我将不胜感激，并会不断修改加以完善。

电子邮件地址：yantaicfl@126.com。

刘彩凤

2021年8月于烟台

目 录

第1章 学习交换冗余技术 ······1

1.1 学习 STP ······2
1.1.1 认识 STP ······2
1.1.2 剖析 STP 的运行 ······3
1.1.3 比较多种类型的 STP ······5
1.1.4 场景一：分析 STP 的基本参数 ······5
1.1.5 场景二：干预 STP 的自动选举 ······8
1.1.6 场景三：实现 STP 负载均衡 ······10
1.1.7 场景四：拓展 STP 的网络影响 ······14
1.1.8 场景五：配置多实例 STP 网络 ······15

1.2 学习链路聚合技术 ······21
1.2.1 认识链路聚合 ······21
1.2.2 场景六：配置二层链路聚合 ······22
1.2.3 场景七：配置三层链路聚合 ······25
1.2.4 场景八：分析链路聚合故障 ······27

1.3 学习网关冗余技术 ······33
1.3.1 认识网关冗余 ······33
1.3.2 剖析网关冗余 ······34
1.3.3 场景九：配置单组 HSRP ······35
1.3.4 场景十：配置多组 HSRP ······37

1.4 挑战练习 ······42

1.5 本章小结 ······43

第2章 学习 LAN 安全技术 ······44

2.1 学习端口安全技术 ······45
2.1.1 认识端口安全 ······45
2.1.2 场景一：配置静态绑定 MAC 地址的端口安全 ······46
2.1.3 场景二：配置粘滞绑定 MAC 地址端口安全 ······48

2.2 学习 IEEE 802.1x 认证技术 ······51
2.2.1 认识 IEEE 802.1x 认证 ······51

2.2.2　场景三：配置 IEEE 802.1x 认证 51
2.3　学习 DHCP 侦听技术 55
2.3.1　认识 DHCP 攻击 55
2.3.2　场景四：配置 DHCP 侦听 56
2.4　学习防止 ARP 攻击技术 59
2.4.1　认识 DAI 的原理及作用 59
2.4.2　场景五：配置 DHCP Snooping 及 DAI 59
2.5　学习 STP 优化技术 61
2.5.1　认识 STP 的优化参数 61
2.5.2　场景六：优化 STP 的网络性能 61
2.6　挑战练习 69
2.7　本章小结 70

第 3 章　学习基础 OSPFv2 71

3.1　学习 OSPFv2 72
3.1.1　学习 OSPFv2 基础知识 72
3.1.2　学习 OSPFv2 的基本特征 73
3.1.3　学习 OSPFv2 的原理 74
3.1.4　学习 OSPFv2 的数据包类型 75
3.1.5　学习 OSPFv2 的运行状态 78
3.1.6　学习 OSPFv2 路由器类型 79
3.2　配置基础 OSPFv2 81
3.2.1　认识 OSPFv2 路由器 ID 81
3.2.2　场景一：配置 OSPFv2 路由器 ID 82
3.2.3　认识单区域 OSPFv2 83
3.2.4　场景二：配置单区域 OSPFv2 83
3.2.5　认识多区域 OSPFv2 86
3.2.6　场景三：配置多区域 OSPFv2 87
3.3　检验 OSPFv2 配置 90
3.3.1　实验一：检验 OSPFv2 配置 90
3.3.2　实验二：查看 OSPFv2 邻居表 92
3.3.3　实验三：查看 OSPFv2 拓扑表 93
3.3.4　实验四：查看 OSPFv2 的 LSA 96
3.3.5　实验五：验证 OSPFv2 路径 102
3.4　干预 OSPFv2 运行 104

		3.4.1 场景四：干预 OSPFv2 邻接关系的建立	104

    3.4.1 场景四：干预 OSPFv2 邻接关系的建立 ··· 104
    3.4.2 场景五：干预 OSPFv2 DR 选举 ·· 110
    3.4.3 场景六：优化 OSPFv2 网络性能 ··· 113
    3.4.4 场景七：干预 OSPFv2 自动选路 ··· 117
3.5 挑战练习 ··· 121
    3.5.1 挑战练习一 ·· 121
    3.5.2 挑战练习二 ·· 122
3.6 本章小结 ··· 123

第 4 章　学习高级 OSPFv2 ·· 124

4.1 完成 OSPFv2 虚链路配置 ··· 125
    4.1.1 认识 OSPFv2 虚链路 ··· 125
    4.1.2 场景一：配置 OSPFv2 虚链路 ··· 126
4.2 完成 OSPFv2 认证配置 ··· 131
    4.2.1 认识 OSPFv2 认证 ··· 131
    4.2.2 场景二：配置 OSPFv2 安全认证 ··· 131
4.3 完成 OSPFv2 路由注入配置 ·· 134
    4.3.1 认识外部路由注入 ·· 134
    4.3.2 场景三：配置外部路由注入及默认路由传播 ··································· 135
    4.3.3 场景四：配置多进程 OSPFv2 ··· 140
4.4 完成 OSPFv2 路由汇总配置 ·· 145
    4.4.1 认识 OSPFv2 路由汇总 ·· 145
    4.4.2 场景五：配置 OSPFv2 域间路由汇总 ·· 146
    4.4.3 场景六：配置 OSPFv2 外部路由汇总 ·· 152
4.5 完成 OSPFv2 特殊区域配置 ·· 155
    4.5.1 认识 OSPFv2 特殊区域 ·· 155
    4.5.2 场景七：配置 OSPFv2 特殊区域 ··· 157
4.6 挑战练习 ··· 169
    4.6.1 挑战练习一 ·· 169
    4.6.2 挑战练习二 ·· 169
4.7 本章小结 ··· 170

第 5 章　学习 ACL 技术 ·· 171

5.1 认识 ACL 技术 ·· 172
    5.1.1 认识 ACL 定义 ·· 172

	5.1.2	学习通配符掩码	172
	5.1.3	认识常用端口号	172
	5.1.4	了解 ACL 应用原则	173
	5.1.5	了解 ACL 应用范围	173
5.2	配置标准 ACL		174
	5.2.1	认识标准 ACL	174
	5.2.2	学习标准 ACL 语法	174
	5.2.3	场景一：配置标准编号 ACL	175
	5.2.4	场景二：配置标准命名 ACL	180
5.3	配置扩展 ACL		182
	5.3.1	认识扩展 ACL	182
	5.3.2	学习扩展 ACL 语法	182
	5.3.3	场景三：配置扩展编号 ACL	183
	5.3.4	场景四：配置扩展命名 ACL	185
5.4	拓展 ACL 应用		188
	5.4.1	认识 VTY 技术	188
	5.4.2	场景五：采用 ACL 限制 VTY 访问	188
	5.4.3	认识 QoS 技术	190
	5.4.4	场景六：将 ACL 应用于 QoS	190
5.5	挑战练习		192
	5.5.1	挑战练习一	192
	5.5.2	挑战练习二	193
5.6	本章小结		194

第 6 章 学习 NAT 技术 195

6.1	认识 NAT 技术		196
	6.1.1	理解 NAT 的定义	196
	6.1.2	熟悉 NAT 的分类	196
	6.1.3	了解 NAT 的术语	196
	6.1.4	理解 NAT 的原理	197
	6.1.5	理解 NAT 的优缺点	200
6.2	配置多种类型的 NAT		200
	6.2.1	认识动态 NAT	200
	6.2.2	场景一：配置动态 NAT	201
	6.2.3	认识动态 PAT	204

		6.2.4 场景二：配置动态 PAT	205
		6.2.5 认识静态 NAT	208
		6.2.6 场景三：配置静态 NAT	208
		6.2.7 认识静态 PAT	212
		6.2.8 场景四：配置静态 PAT	213
		6.2.9 常见 NAT 故障排错命令	219
	6.3	配置拓展 NAT	219
		6.3.1 场景五：在无线路由器上完成 NAT 内置配置	219
		6.3.2 场景六：配置由外至内映射的 NAT	222
	6.4	挑战练习	225
		6.4.1 挑战练习一	225
		6.4.2 挑战练习二	226
	6.5	本章小结	228

第 7 章 学习基础 IPv6 — 229

7.1	学习 IPv6 基础知识	230
	7.1.1 学习 IPv6 地址	230
	7.1.2 认识 ICMPv6	231
	7.1.3 认识 NDP	231
7.2	配置 IPv6 地址	232
	7.2.1 场景一：配置 IPv6 全局单播地址	232
	7.2.2 场景二：配置 IPv6 的 EUI-64 地址	235
	7.2.3 场景三：配置 IPv6 链路本地地址	238
	7.2.4 场景四：配置 IPv6 任播地址	240
7.3	配置 SLAAC 与 DHCPv6	247
	7.3.1 认识 SLAAC 与 DHCPv6	247
	7.3.2 学习 DHCPv6 配置语法	247
	7.3.3 场景五：完成 SLAAC（无状态地址自动配置）	248
	7.3.4 场景六：配置无状态 DHCPv6 服务	251
	7.3.5 场景七：配置状态化 DHCPv6 服务	253
7.4	挑战练习	256
	7.4.1 挑战练习一	256
	7.4.2 挑战练习二	257
7.5	本章小结	258

第 8 章　学习高级 IPv6 ··· 259

8.1　学习 IPv6 静态路由 ··· 260
8.1.1　认识 IPv6 静态路由 ·· 260
8.1.2　IPv6 静态路由分类 ·· 260
8.1.3　IPv6 静态路由语法 ·· 260
8.1.4　学习 IPv6 静态路由配置方法 ·· 261
8.1.5　场景一：配置 IPv6 静态路由 ·· 261

8.2　学习 IPv6 动态路由 ··· 265
8.2.1　认识 RIPng 路由协议 ··· 265
8.2.2　学习 RIPng 配置语法 ··· 266
8.2.3　场景二：配置 RIPng 动态路由 ·· 266
8.2.4　认识 OSPFv3 路由协议 ··· 269
8.2.5　学习 OSPFv3 配置语法 ··· 270
8.2.6　场景三：配置 OSPFv3 动态路由 ·· 270

8.3　学习 IPv6 ACL ··· 274
8.3.1　认识 IPv6 ACL ·· 275
8.3.2　学习 IPv6 ACL 语法 ·· 275
8.3.3　场景四：配置 IPv6 ACL ··· 275

8.4　配置 IPv6 综合案例 ··· 280
8.5　挑战练习 ··· 287
8.5.1　挑战练习一 ··· 288
8.5.2　挑战练习二 ··· 288
8.6　本章小结 ··· 289

第1章 >>>

学习交换冗余技术

本章要点：

- 学习 STP
- 学习链路聚合技术
- 学习网关冗余技术
- 挑战练习
- 本章小结

本章介绍了如何采用交换冗余技术，确保冗余交换网络运行的可靠性与稳定性。其中，1.1 节介绍 STP，讲解如何采用 STP 将物理冗余的环形网络构建成逻辑树状网络；1.2 节介绍链路聚合技术，讲解在网络带宽不足时如何将多条物理链路捆绑为一条逻辑链路；1.3 节介绍网关冗余技术，讲解在网络中如果出现网关宕机情况，如何无缝地切换至备用网关。为使读者更加透彻地学习与理解交换冗余技术，笔者在本章中设计了 10 个应用场景和 1 个挑战练习来检验读者对本章知识的理解与应用，避免枯燥的强化记忆。

1.1 学习 STP

1.1.1 认识 STP

1. STP 的定义

STP（Spanning Tree Protocol，生成树协议）是二层链路管理协议，其主要功能是在保证网络中没有环路的基础上，在第二层链路中提供冗余路径，以保证网络可靠、稳定地运行。

2. 二层环路可能导致的问题

① 交换机 CAM 表不稳定：在交换机的不同端口收到同一个帧的多个副本，导致 MAC 地址表不稳定。

② 广播风暴：如果没有环路避免机制，交换机可能无休止地进行广播泛洪。

③ 帧的多重传输：将单播帧的多个副本传输到目标工作站。

3. STP 的参数

① 网桥 ID：
- 网桥 ID（8 字节）=网桥优先级（2 字节）+网桥 MAC 地址（6 字节）；
- 默认优先级为 32768，范围为 0~65535，值越小优先级越高。

② 端口 ID：
- 端口 ID（2 字节）=端口优先级（1 字节）+端口 ID（1 字节）；
- 默认优先级为 128，范围为 0~255，值越小优先级越高。

③ 路径开销：
- 交换机到达根桥交换机的总开销，值越小优先级越高，与端口带宽有关；
- 路径开销如表 1-1 所示。

表 1-1　路径开销

带　宽	开销（Cost）
10 Gbps	2
1 Gbps	4
100 Mbps	19
10 Mbps	100

4．STP 的计时器

① 从阻塞到转发状态通常要 30～50 s（默认为 50 s，即 20 s+15 s+15 s）。

② Hello 时间：根网桥发送配置 BPDU（Bridge Protocol Data Unit，网桥协议数据单元）报文的时间间隔（2 s）。

③ 转发延迟时间：从侦听（Listening）状态到学习（Learning）状态，或者从学习状态到转发（Forwarding）状态所需要的时间（15 s）。

④ 最大存活期：在丢弃 BPDU 报文之前，网桥存储 BPDU 报文的时间（20 s），如果在 20 s 内没有连续收到 10 个 BPDU 报文，则网桥进入侦听状态。

5．STP 的增强特性

① PortFast：端口一旦连接了设备，端口可绕过 Listening 状态和 Learning 状态直接进入 Forwarding 状态。

② BPDU Guard：在端口模式下，端口收到 BPDU 报文后将立即切换到 err-disable 状态。

③ UplinkFast：配置 UplinkFast 的交换机为末梢交换机，能够在直连链路发生故障后提供快速收敛功能。

④ Root Guard：防止新加入交换机抢占根角色，导致生成树重新选择，影响网络的稳定性。

⑤ BackbooneFast：只对直连交换机所发生的故障进行快速响应，减少默认的收敛时间。

1.1.2　剖析 STP 的运行

图 1-1 所示为在 3 台交换机上运行 STP。

1．STP 选举过程

（1）选举一个根桥交换机

广播域中的所有交换机都会参与选举，网桥 ID 最小的交换机被选为生成树的根桥交换机。在图 1-1 中，由于优先级相同，所以 MAC 地址小的交换机为根桥交换机，即 Switch3 为根桥交换机。

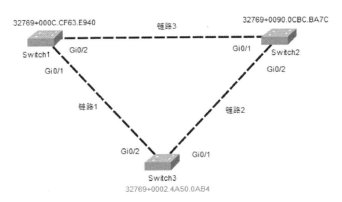

图 1-1　在 3 台交换机上运行 STP

（2）在所有非根桥交换机上选举根端口

STP 在每个非根网桥上选举 1 个根端口，该端口所连接的路径一定是该交换机到根网桥开销最小的路径。选举根端口次序是：Cost>对端网桥 ID>对端端口 ID。因为 Switch1 的 Gi0/1 端口到 Switch3 的 Cost 为 4，Switch1 的 Gi0/2 端口到 Switch3 的 Cost 为 8，所以 Switch1 的 Gi0/1 端口为根端口。以此类推，Switch2 的 Gi0/2 端口为根端口。

（3）在每条链路上选举指定端口

STP 为每条链路选举一个指定端口，每条链路只能有一个指定端口。选举次序是：Cost>网桥 ID>端口 ID。因为 Switch3 为根桥交换机，Switch3 的所有端口到根交换机的 Cost 都为 0，所以 Switch3 的所有端口都是指定端口；在链路 3 上，Switch1 的 Gi0/2 端口和 Switch2 的 Gi0/1 端口到 Switch3 的 Cost 相同，Switch1 的网桥 ID 比 Switch2 的网桥 ID 小，Switch1 的 Gi0/2 端口为链路 3 上的指定端口。

（4）所有其他端口都为非指定端口

除根端口和指定端口之外的其他所有交换机端口都是非指定端口。例如，Switch2 的 Gi0/1 端口为非指定端口。

2．STP 端口角色

① 根端口：最靠近根网桥的交换机端口，每台非根交换机上都会选择一个根端口。

② 指定端口：网络中允许转发流量的非根端口。在每条链路上都会选择一个指定端口，根交换机的所有端口都是指定端口。

③ 非指定端口：除根端口和指定端口之外的其余端口都将成为非指定端口，非指定端口只能接收流量，但不能转发流量。

3．STP 端口状态

STP 端口状态如表 1-2 所示。

表 1-2　STP 端口状态

状　态	说　明
Disable（禁用）	不收发任何报文
Blocking（阻塞）	不接收也不转发数据帧，接收但不转发 BPDU 报文，不学习 MAC 地址
Listening（侦听）	不接收也不转发数据帧，接收并且发送 BPDU 报文，不学习 MAC 地址
Learning（学习）	不接收也不转发数据帧，接收并且发送 BPDU 报文，学习 MAC 地址
Forwarding（转发）	接收并转发数据帧，接收并且发送 BPDU 报文，学习 MAC 地址

4．STP 端口状态转换过程

STP 端口状态转换过程如图 1-2 所示。

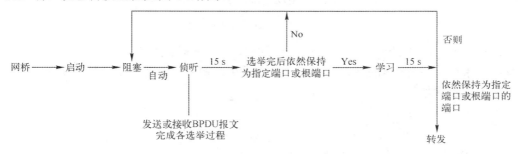

图 1-2　STP 端口状态转换过程

1.1.3　比较多种类型的 STP

如表 1-3 所示，比较多种 STP 的类型。

表 1-3　比较多种 STP 的类型

协　议	标　准	资源需求	收敛速度	生成树计算
STP	IEEE 802.1d	低	慢	所有 VLAN 一棵树
PVST+	思科专有	高	慢	每个 VLAN 一棵树
RSTP	IEEE 802.1w	中	快	所有 VLAN 一棵树
快速 PVST+	思科专有	极高	快	每个 VLAN 一棵树
MSTP（MST）	IEEE 802.1s	中高	快	每个实例一棵树

1.1.4　场景一：分析 STP 的基本参数

3 台交换机 Switch1、Switch2 和 Switch3 通过 3 条链路互连，在交换机之间组成一个环，通过 PVST 生成树选举出根端口和指定端口；阻塞全部非指定端口，使有环路的物理网络变成无环路的逻辑网络，实现冗余；如图 1-3 所示，分析 STP 的基本参数。

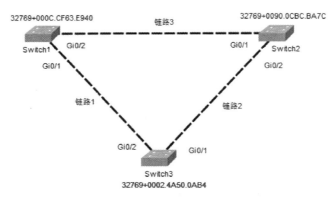

图 1-3　分析 STP 的基本参数

注意：在 Cisco Gatalyst 交换机上默认开启 PVST 生成树。

任务需求：使用命令 **show spanning-tree** 查看生成树协议并找出交换机端口状态。

具体配置步骤如下。

步骤一：查看生成树协议

（1）在 Switch1 上查看生成树协议

```
Switch1#show spanning-tree
VLAN0001
  Spanning tree enabled protocol ieee
  Root ID    Priority    32769                          //根交换机优先级为 32769（32768+1）
             Address     0002.4A50.0AB4                 //根交换机的 MAC 地址
             Cost        4                              //根端口到根交换机的 Cost
             Port        25(GigabitEthernet0/1)         //根端口的端口编号
             Hello Time  2 sec   Max Age 20 sec   Forward Delay 15 sec

  Bridge ID  Priority    32769   (priority 32768 sys-id-ext 1)    //本交换机的优先级
             Address     000C.CF63.E940                 //本交换机的 MAC 地址
             Hello Time 2 sec   Max Age 20 sec   Forward Delay 15 sec
                                                        //交换机 Hello 时间、最大存活时间和转发延迟时间
             Aging Time 20

Interface           Role Sts     Cost      Prio.Nbr Type
---------------- ---- --- --------- --------  --------------------------------
```

Gi0/1		Root FWD	4	128.25	P2p
				//端口角色为根端口，端口状态为转发	
Gi0/2		Desg FWD	4	128.26	P2p
				//端口角色为指定端口，端口状态为转发	

（2）在 Switch2 上查看生成树协议

```
Switch2#show spanning-tree
VLAN0001
  Spanning tree enabled protocol ieee
  Root ID    Priority    32769
             Address     0002.4A50.0AB4
             Cost        4
             Port        26(GigabitEthernet0/2)
             Hello Time  2 sec   Max Age 20 sec   Forward Delay 15 sec

  Bridge ID  Priority    32769   (priority 32768 sys-id-ext 1)
             Address     0090.0CBC.BA7C
             Hello Time  2 sec   Max Age 20 sec   Forward Delay 15 sec
             Aging Time  20

Interface        Role Sts     Cost      Prio.Nbr Type
---------------- ---- ---  ---------  --------  --------------------------------
Gi0/2            Root FWD     4         128.26   P2p
Gi0/1            Altn BLK     4         128.25   P2p
```
                                   //端口角色为替代端口，端口状态为阻塞

（3）在 Switch3 上查看生成树协议

```
Switch3#show spanning-tree
VLAN0001
  Spanning tree enabled protocol ieee
  Root ID    Priority    32769
             Address     0002.4A50.0AB4
             This bridge is the root            //这是根交换机
             Hello Time  2 sec   Max Age 20 sec   Forward Delay 15 sec

  Bridge ID  Priority    32769   (priority 32768 sys-id-ext 1)
             Address     0002.4A50.0AB4
```

```
                    Hello Time 2 sec   Max Age 20 sec   Forward Delay 15 sec
                    Aging Time 20

   Interface        Role Sts        Cost        Prio.Nbr Type
   ---------------- ---- ---        ----        ---------------
   Gi0/1            Desg FWD         4          128.25   P2p
   Gi0/2            Desg FWD         4          128.26   P2p
```

1.1.5 场景二：干预 STP 的自动选举

将 4 台交换机 S1、S2、S3 和 S4 组成一个环，通过修改 PVST 的优先级来干预 PVST 根选举，使性能更强的交换机 S1 被选为根交换机，干预 STP 选举拓扑，如图 1-4 所示。

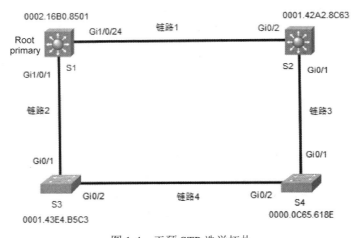

图 1-4 干预 STP 选举拓扑

按照默认配置，默认优先级是相同的（默认优先级为 32768），交换机 S4 为根交换机。但是，S4 是老旧设备，无法担任根交换机这个角色，我们将通过修改设备优先级的方式让最新购买的交换机 S1 成为根交换机。

修改优先级的语法如下：

```
spanning-tree vlan vlan-id priority bridge-priority
```

具体配置步骤如下所述。

步骤一：修改交换机 S1 的优先级

```
S1(config)#spanning-tree vlan 1 priority 4096
```

步骤二：查看交换机 S1 的生成树

```
S1#show spanning-tree
VLAN0001
  Spanning tree enabled protocol ieee
  Root ID    Priority    4097
             Address     0002.16B0.8501
             This bridge is the root
             Hello Time  2 sec   Max Age 20 sec   Forward Delay 15 sec

  Bridge ID  Priority    4097    (priority 4096 sys-id-ext 1)
             Address     0002.16B0.8501
             Hello Time 2 sec   Max Age 20 sec   Forward Delay 15 sec
             Aging Time20

Interface        Role Sts     Cost      Prio.Nbr Type
---------------- ---- ---  ---------   --------------------------------
Gi1/0/24         Desg FWD     4         128.24   P2p
Gi1/0/1          Desg LRN     4         128.1    P2p
```

通过一番选举之后,管理员发现链路 4 被阻塞了。因为 S3 与 S4 之间有业务通信,所以不能阻塞,只能阻塞链路 3。通过修改 S3 的优先级,让 S4 的 Gi0/2 端口被选举为根端口。

步骤三:修改交换机 S3 的优先级

```
S3(config)#spanning-tree vlan 1 priority 8192
```

步骤四:查看交换机 S4 交换机的端口变化情况

```
S4#show spanning-tree
VLAN0001
  Spanning tree enabled protocol ieee
  Root ID    Priority    4097
             Address     0002.16B0.8501
             Cost        8
             Port        26(GigabitEthernet0/2)
             Hello Time  2 sec   Max Age 20 sec   Forward Delay 15 sec

  Bridge ID  Priority    32769   (priority 32768 sys-id-ext 1)
             Address     0000.0C65.618E
             Hello Time 2 sec   Max Age 20 sec   Forward Delay 15 sec
             Aging Time20
```

```
Interface         Role Sts      Cost      Prio.Nbr    Type
---------------- ---- ---      ------    --------    ----
Gi0/2             Root FWD      4         128.26      P2p
Gi0/1             Altn BLK      4         128.25      P2p
```

通过修改交换机 S3 的优先级，使交换机 S4 的 Gi0/1 端口被阻塞。

步骤五：修改交换机 S4 的 Cost 值

```
S4(config)# interface GigabitEthernet 0/1
S4(config-if)# spanning-tree vlan 1 cost 1
```

步骤六：查看交换机 S4 的端口变化情况

```
S4#show spanning-tree
VLAN0001
  Spanning tree enabled protocol ieee
  Root ID    Priority    4097
             Address     0002.16B0.8501
             Cost        5                              //到根交换机的 Cost（4+1）
             Port        25(GigabitEthernet0/1)
             Hello Time  2 sec   Max Age 20 sec   Forward Delay 15 sec

  Bridge ID  Priority    32769  (priority 32768 sys-id-ext 1)
             Address     0000.0C65.618E
             Hello Time 2 sec   Max Age 20 sec   Forward Delay 15 sec
             Aging Time  20

Interface         Role Sts      Cost      Prio.Nbr    Type
---------------- ---- ---      ------    --------    ----
Gi0/2             Altn BLK      4         128.26      P2p
Gi0/1             Root FWD      1         128.25      P2p
```

通过修改 Gi0/1 端口的 Cost 来修改交换机的根端口。

1.1.6 场景三：实现 STP 负载均衡

通过对 4 台交换机进行配置，使不同的 VLAN 数据分别从不同设备转发出去，其中，VLAN 10 和 VLAN 20 的数据优先从交换机 CS1 转发，VLAN 30 和 VLAN 40 的数据优先从交换机 CS2 转发。实现 STP 负载均衡拓扑，如图 1-5 所示。

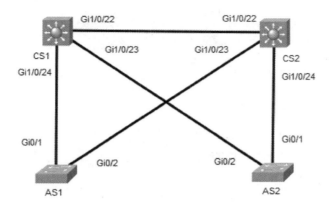

图 1-5　实现 STP 负载均衡拓扑

通过修改不同 VLAN 的网桥 ID 优先级来影响不同 VLAN 的数据流向。具体配置步骤如下所述。

步骤一：在 4 台交换机上创建 4 个 VLAN

（1）在 AS1 上创建 VLAN

```
AS1(config)#vlan 10
AS1(config-vlan)# vlan 20
AS1(config-vlan)# vlan 30
AS1(config-vlan)# vlan 40
```

（2）在 AS2 上创建 VLAN

```
AS2(config)#vlan 10
AS2(config-vlan)# vlan 20
AS2(config-vlan)# vlan 30
AS2(config-vlan)# vlan 40
```

（3）在 CS1 上创建 VLAN

```
CS1(config)#vlan 10
CS1(config-vlan)# vlan 20
CS1(config-vlan)# vlan 30
CS1(config-vlan)# vlan 40
```

（4）在 CS2 上创建 VLAN

```
CS2(config)#vlan 10
CS2(config-vlan)# vlan 20
```

```
CS2(config-vlan)# vlan 30
CS2(config-vlan)# vlan 40
```

步骤二：设置交换机 trunk 端口

（1）设置交换机 AS1 的 trunk 端口

```
AS1(config)#interface range GigabitEthernet 0/1-2
AS1(config-if-range)#switchport mode trunk
```

（2）设置交换机 AS2 的 trunk 端口

```
AS2(config)#interface range GigabitEthernet 0/1-2
AS2(config-if-range)#switchport mode trunk
```

（3）设置交换机 CS1 的 trunk 端口

```
CS1(config)#interface range GigabitEthernet 1/0/22-24
CS1(config-if-range)#switchport trunk encapsulation dot1q
CS1(config-if-range)#switchport mode trunk
```

（4）设置交换机 CS2 的 trunk 端口

```
CS2(config)#interface range GigabitEthernet 1/0/22-24
CS2(config-if-range)#switchport trunk encapsulation dot1q
CS2(config-if-range)#switchport mode trunk
```

步骤三：通过控制不同 VLAN 的根来影响数据转发

- 交换机 CS1 为 VLAN 10、VLAN 20 的主根，VLAN 30、VLAN 40 的次根；
- 交换机 CS2 为 VLAN 10、VLAN 20 的次根，VLAN 30、VLAN 40 的主根。

（1）在交换机 CS1 上配置生成树

```
CS1(config)#spanning-tree vlan 10 root primary
CS1(config)#spanning-tree vlan 20 root primary
CS1(config)#spanning-tree vlan 30 root secondary
CS1(config)#spanning-tree vlan 40 root secondary
```

（2）在交换机 CS2 上配置生成树

```
CS2(config)#spanning-tree vlan 10 root secondary
CS2(config)#spanning-tree vlan 20 root secondary
CS2(config)#spanning-tree vlan 30 root primary
```

```
CS2(config)#spanning-tree vlan 40 root primary
```

步骤四：在 CS1 上查看 VLAN 10 的生成树状态

```
CS1#show spanning-tree vlan 10
VLAN0010
  Spanning tree enabled protocol ieee
  Root ID    Priority    24586
             Address     000D.BDDB.A559
             This bridge is the root
             Hello Time  2 sec  Max Age 20 sec  Forward Delay 15 sec

  Bridge ID  Priority    24586   (priority 24576 sys-id-ext 10)
             Address     000D.BDDB.A559
             Hello Time 2 sec   Max Age 20 sec  Forward Delay 15 sec
             Aging Time 20

Interface        Role Sts    Cost        Prio.Nbr   Type
---------------- ---- ---    ---------   --------   --------------------------
Gi1/0/22         Desg FWD    4           128.22     P2p
Gi1/0/23         Desg FWD    4           128.23     P2p
Gi1/0/24         Desg FWD    4           128.24     P2p
```

步骤五：在 CS2 上查看 VLAN 30 的生成树状态

```
CS2#show spanning-tree vlan 30
VLAN0030
  Spanning tree enabled protocol ieee
  Root ID    Priority    24606
             Address     00D0.BA2B.1452
             This bridge is the root
             Hello Time  2 sec  Max Age 20 sec  Forward Delay 15 sec

  Bridge ID  Priority    24606   (priority 24576 sys-id-ext 30)
             Address     00D0.BA2B.1452
             Hello Time 2 sec   Max Age 20 sec  Forward Delay 15 sec
             Aging Time 20

Interface         Role Sts     Cost         Prio.Nbr    Type
```

```
----------------- ---- --- --------- --------- ---------
Gi1/0/22           Desg FWD   4       128.22    P2p
Gi1/0/23           Desg FWD   4       128.23    P2p
Gi1/0/24           Desg FWD   4       128.24    P2p
```

从步骤四和步骤五上可以看出来，VLAN 10 的根交换机为 CS1，VLAN 30 的根交换机为 CS2。同理，查看 VLAN 20 和 VLAN 40 的生成树状态。

思考与观察：请读者自己在 CS1 上设置故障，观察 VLAN 10 和 VLAN 20 生成树根的自动切换情况，进一步理解生成树次根的作用及冗余的好处；当故障排除后，VLAN 10 和 VLAN 20 生成树的根又恢复为初始设计。

1.1.7 场景四：拓展 STP 的网络影响

交换机 S1 通过 HUB 与自身两个端口相连（见图 1-6），请查看交换机 S1 如何指定其端口；如果交换机 S1 通过 HUB 与自身多个端口相连，请查看情况又如何。

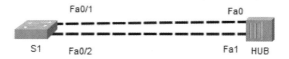

图 1-6　交换机 S1 通过 HUB 与自身两个端口相连

将一台交换机的两个端口连接到 HUB，就相当于交换机的两个端口连到一起。

如图 1-6 所示，在交换机 S1 上查看生成树，具体配置步骤如下所述。

```
S1 #show spanning-tree
VLAN0001
  Spanning tree enabled protocol ieee
  Root ID    Priority    32769
             Address     0002.4A44.B9C4
             This bridge is the root
             Hello Time  2 sec   Max Age 20 sec   Forward Delay 15 sec

  Bridge ID  Priority    32769   (priority 32768 sys-id-ext 1)
             Address     0002.4A44.B9C4
             Hello Time 2 sec   Max Age 20 sec   Forward Delay 15 sec
             Aging Time  20

Interface        Role Sts   Cost      Prio.Nbr   Type
---------------- ---- ---   --------  ---------  ---------
Fa0/1            Desg FWD   19        128.1      Shr
```

| Fa0/2 | Altn BLK | 19 | 128.2 | Shr |

如图 1-7 所示，交换机 S1 通过 HUB 与自身多个端口相连，在交换机 S1 上查看生成树，具体配置步骤如下所述。

图 1-7 交换机 S1 通过 HUB 与自身多个端口相连

```
S1#show spanning-tree
VLAN0001
  Spanning tree enabled protocol ieee
  Root ID    Priority    32769
             Address     0002.4A44.B9C4
             This bridge is the root
             Hello Time  2 sec  Max Age 20 sec  Forward Delay 15 sec

  Bridge ID  Priority    32769   (priority 32768 sys-id-ext 1)
             Address     0002.4A44.B9C4
             Hello Time  2 sec  Max Age 20 sec  Forward Delay 15 sec
             Aging Time  20

Interface        Role Sts       Cost      Prio.Nbr  Type
---------------- ---- ---       --------- --------  --------------------------------
Fa0/3            Altn BLK       19        128.3     Shr
Fa0/4            Altn BLK       19        128.4     Shr
Fa0/1            Desg FWD       19        128.1     Shr
Fa0/2            Altn BLK       19        128.2     Shr
```

由上可知，只一个交换机端口处于转发状态，其余端口都被阻塞了。

该场景拓展了 STP 的网络影响。

1.1.8 场景五：配置多实例 STP 网络

MSTP 能够兼容 STP 和 RSTP，而且能将一个交换网络划分成多个域，每个域内形成多棵生成树，每棵生成树之间彼此相互独立，将环形网络修剪为树形，避免报文在环路中的增生和无限循环，同时还可以提供数据转发的冗余路径，在数据转发过程中实现 VLAN 数据的负载均衡。MSTP 在实际交换网络中被应用广泛，但超出了 Packet Tracer 的支持范围，为了使读者掌握该技术的应用，本场景采用 eve-ng 模拟器进行了补充实验。

在如图 1-8 所示的 MSTP 网络拓扑中，4 台交换机 S1、S2、S3 和 S4 分别相连，组成很多个环形结构,其中,交换机 S1 和 S2 为核心层交换机,交换机 S3 和 S4 为接入层交换机,可通过配置 MSTP 来使数据分流并防止产生环路,保证数据正常转发。

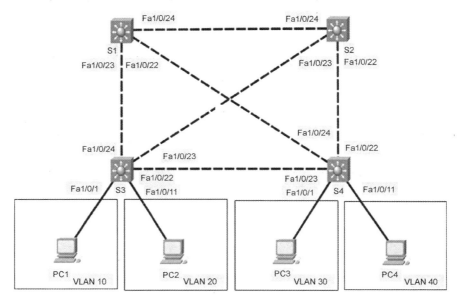

图 1-8 MSTP 网络拓扑

交换机 S1 和 S2 为核心层交换机,交换机 S3 和 S4 为接入层交换机,在 S1、S2、S3、S4 上运行 MSTP 以防止二层环路,VLAN 10 和 VLAN 20 的数据流量经 S1 转发,当 S1 失效时经 S2 转发;VLAN 30 和 VLAN 40 的数据流量经 S2 转发,当 S2 失效时经 S1 转发。

配置 MSTP 的语法如下:

spanning-tree mode mstp	//修改生成树为 MSTP
spanning-tree mstp configuration	//进入 MSTP 配置模式
revision 1	//定义修订版本号
instance 1 vlan *vlan-id*	//将 vlan 划分到实例中
spanning-tree mstp 1 priority 4096	//修改实例 1 的优先级

具体配置步骤如下所述。

步骤一:创建 VLAN 并将端口添加到 VLAN 中

(1)在交换机 S1 上创建 VLAN

```
S1(config)#vlan 10
S1(config-vlan)#vlan 20
S1(config-vlan)#vlan 30
S1(config-vlan)#vlan 40
```

（2）在交换机 S2 上创建 VLAN

```
S2(config)#vlan 10
S2(config-vlan)#vlan 20
S2(config-vlan)#vlan 30
S2(config-vlan)#vlan 40
```

（3）在交换机 S3 上创建 VLAN 并划分端口

```
S3(config)#vlan 10
S3(config-vlan)#vlan 20
S3(config)#interface range fastEthernet 1/0/1 -10
S3(config-if-range)#switchport mode access
S3(config-if-range)#switchport access vlan 10
S3(config-if-range)#exit
S3(config)#interface range fastEthernet 1/0/11 -20
S3(config-if-range)#switchport mode access
S3(config-if-range)#switchport access vlan 20
```

（4）在交换机 S4 上创建 VLAN 并划分端口

```
S4(config)#vlan 30
S4(config-vlan)#vlan 40
S4(config)#interface range fastEthernet 1/0/1 -10
S4(config-if-range)#switchport mode access
S4(config-if-range)#switchport access vlan 30
S4(config-if-range)#exit
S4(config)#interface range fastEthernet 1/0/11 -20
S4(config-if-range)#switchport mode access
S4(config-if-range)#switchport access vlan 40
```

步骤二：配置链路端口类型为 trunk

（1）在交换机 S1 上配置 trunk 端口

```
S1(config)#interface range fastEthernet 1/0/22 -24
S1(config-if-range)#switchport trunk encapsulation dot1q
S1(config-if-range)#switchport mode trunk
```

（2）在交换机 S2 上配置 trunk 端口

```
S2(config)#interface range fastEthernet 1/0/22 -24
```

```
S2(config-if-range)#switchport trunk encapsulation dot1q
S2(config-if-range)#switchport mode trunk
```

（3）在交换机 S3 上配置 trunk 端口

```
S3(config)#interface range fastEthernet 1/0/22 -24
S3(config-if-range)#switchport trunk encapsulation dot1q
S3(config-if-range)#switchport mode trunk
```

（4）在交换机 S4 上配置 trunk 端口

```
S4(config)#interface range fastEthernet 1/0/22 -24
S4(config-if-range)#switchport trunk encapsulation dot1q
S4(config-if-range)#switchport mode trunk
```

步骤三：修改生成树模式

（1）修改交换机 S1 的生成树模式

```
S1(config)#spanning-tree mode mst
```

（2）修改交换机 S2 的生成树模式

```
S2(config)#spanning-tree mode mst
```

（3）修改交换机 S3 的生成树模式

```
S3(config)#spanning-tree mode mst
```

（4）修改交换机 S4 的生成树模式

```
S4(config)#spanning-tree mode mst
```

步骤四：创建生成树实例

（1）在交换机 S1 上创建实例

```
S1(config)#spanning-tree mst configuration
S1(config-mst)#instance 1 vlan 10,20
S1(config-mst)#instance 2 vlan 30,40
```

（2）在交换机 S2 上创建实例

```
S2(config)#spanning-tree mst configuration
S2(config-mst)#instance 1 vlan 10,20
```

S2(config-mst)#**instance 2 vlan 30,40**

（3）在交换机 S3 上创建实例

S3(config)#**spanning-tree mst configuration**
S3(config-mst)#**instance 1 vlan 10,20**
S3(config-mst)#**instance 2 vlan 30,40**

（4）在交换机 S4 上创建实例

S4(config)#**spanning-tree mst configuration**
S4(config-mst)#**instance 1 vlan 10,20**
S4(config-mst)#**instance 2 vlan 30,40**

步骤五：修改实例优先级

（1）在交换机 S1 上修改不同实例的优先级

S1(config)#**spanning-tree mst 1 priority 4096**
S1(config)#**spanning-tree mst 2 priority 8192**

（2）在交换机 S2 上修改不同实例的优先级

S2(config)#**spanning-tree mst 1 priority 8192**
S2(config)#**spanning-tree mst 2 priority 4096**

步骤六：在交换机 S1 上查看生成树状态

```
S1#show spanning-tree mst

##### MST0      vlans mapped:    1-9,11-19,21-29,31-39,41-4094
Bridge          address 04fe.7f30.6200    priority      32768 (32768 sysid 0)
Root            this switch for the CIST
Operational     hello time 2 , forward delay 15, max age 20, txholdcount 6
Configured      hello time 2 , forward delay 15, max age 20, max hops      20

Interface       Role Sts     Cost        Prio.Nbr  Type
---------------- ---- --------- -------- --------------------
Fa1/0/22        Desg FWD     200000      128.111   P2p
Fa1/0/23        Desg FWD     200000      128.112   P2p
Fa1/0/24        Desg FWD     200000      128.113   P2p

##### MST1      vlans mapped:    10,20
```

```
    Bridge         address 04fe.7f30.6200   priority     4097   (4096 sysid 1)
    Root           this switch for MST1

    Interface        Role Sts      Cost        Prio.Nbr  Type
    ---------------- ---- ---   ---------    --------- --------------------------------
    Fa1/0/22         Desg FWD     200000       128.111   P2p
    Fa1/0/23         Desg FWD     200000       128.112   P2p
    Fa1/0/24         Desg FWD     200000       128.113   P2p

    ##### MST2      vlans mapped:   30,40
    Bridge         address 04fe.7f30.6200   priority     8194   (8192 sysid 2)
    Root           address 04fe.7f30.8380   priority     4098   (4096 sysid 2)
                   port      Fa3/0/1        cost         200000       rem hops 19

    Interface        Role Sts      Cost        Prio.Nbr  Type
    ---------------- ---- ---   ---------    --------- --------------------------------
    Fa1/0/22         Root FWD     200000       128.111   P2p
    Fa1/0/23         Desg FWD     200000       128.112   P2p
    Fa1/0/24         Desg FWD     200000       128.113   P2p
```

步骤七：在交换机 S2 上查看生成树状态

```
S2#show spanning-tree mst

##### MST0      vlans mapped:   1-9,11-19,21-29,31-39,41-4094
Bridge         address 04fe.7f30.8380   priority     32768 (32768 sysid 0)
Root           address 04fe.7f30.6200   priority     32768 (32768 sysid 0)
               port      Fa2/0/1        path cost     0
Regional Root address 04fe.7f30.6200   priority     32768 (32768 sysid 0)
                                                     internal cost 200000     rem hops 19
Operational    hello time 2 , forward delay 15, max age 20, txholdcount 6
Configured     hello time 2 , forward delay 15, max age 20, max hops   20

Interface        Role Sts      Cost        Prio.Nbr Type
---------------- ---- ---   ---------    --------- --------------------------------
Fa1/0/22         Root FWD     200000       128.57    P2p
Fa1/0/23         Altn BLK     200000       128.58    P2p
Fa1/0/24         Desg FWD     200000       128.59    P2p

##### MST1      vlans mapped:   10,20
```

```
Bridge        address 04fe.7f30.8380   priority    8193    (8192 sysid 1)
Root          address 04fe.7f30.6200   priority    4097    (4096 sysid 1)
              port    Fa2/0/1          cost        200000      rem hops 19

Interface       Role Sts     Cost        Prio.Nbr    Type
--------------- ---- ---     ---------   --------    ---------------------
Fa1/0/22        Root FWD     200000      128.57      P2p
Fa1/0/23        Desg FWD     200000      128.58      P2p
Fa1/0/24        Desg FWD     200000      128.59      P2p

##### MST2      vlans mapped:   30,40
Bridge        address 04fe.7f30.8380   priority    4098    (4096 sysid 2)
Root          this switch for MST2

Interface       Role Sts     Cost        Prio.Nbr    Type
--------------- ---- ---     ---------   --------    ---------------------
Fa1/0/22        Desg FWD     200000      128.57      P2p
Fa1/0/23        Desg FWD     200000      128.58      P2p
Fa1/0/24        Desg FWD     200000      128.59      P2p
```

1.2 学习链路聚合技术

1.2.1 认识链路聚合

链路聚合是一种将交换机之间的多条链路进行捆绑的技术，思科交换机之间的链路捆绑被称为 EtherChannel。

1．EtherChannel 的功能

EtherChannel 可将多条物理链路捆绑成一条逻辑链路来增加带宽，提供链路冗余。

2．EtherChannel 的优点

- 增加带宽；
- 提供链路冗余；
- 避免二层环路；
- 实现负载分担。

3. EtherChannel 支持的协议

① PAgP：Port Aggregation Protocol，端口汇聚协议。该协议是 Cisco 私有协议，PAgP 数据包每 30 s 发送一次。

② LACP：Link Aggregation Control Protocol，链路汇聚控制协议。该协议是 IEEE 标准协议，工作方式与 PAgP 类似。

PAgP 和 LACP 的模式如表 1-4 所示。

表 1-4 PAgP 和 LACP 的模式

模 式	描 述
auto	在该模式下，端口会对 PAgP 数据包做出响应，但不主动发起协商
desirable	在该模式下，端口会发送 PAgP 数据包来主动与其他端口进行协商
on	在该模式下不使用 PAgP 或 LACP，而是强制端口与邻居形成 Etherchannel
passive	被动协商模式，端口对 LACP 数据包做出响应，但不主动发起协商
active	主动协商模式，端口主动发送 LACP 数据包与其他端口进行协商

1.2.2 场景六：配置二层链路聚合

如图 1-9 所示，交换机 AS1 与 AS2 单条链路互连带宽不够，采用两条链路连接可增加网络带宽，但是 STP 会阻塞其中一条链路。为了解决此问题，可采用链路聚合的方式使两条链路合并成一条链路，以增加网络带宽，实现二层链路聚合。

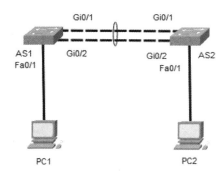

图 1-9 二层链路聚合

配置静态二层链路聚合的语法如下：

```
channel-protocol {lacp | pagp}                                    //选择链路聚合协议
channel-group channel-number mode {active | auto | desirable | on passive}
                                                                  //配置链路聚合状态
port-channel load-balance {dst-mac |src-mac | src-dst-mac | dst-ip | src-ip | src-dst-ip }
                                                                  //配置负载均衡
```

具体配置步骤如下所述。

步骤一：配置端口所属通道组和工作模式

（1）在交换机 AS1 上配置端口

```
AS1(config)#interface range GigabitEthernet 0/1-2
AS1(config-if-range)#channel-group 1 mode on
```

（2）在交换机 AS2 上配置端口

```
AS2(config)#interface range GigabitEthernet 0/1-2
AS2(config-if-range)#channel-group 1 mode on
```

步骤二：配置聚合端口

（1）在交换机 AS1 上配置聚合端口

```
AS1(config)#interface port-channel 1
AS1(config-if)#switchport mode trunk
```

（2）在交换机 AS2 上配置聚合端口

```
AS2(config)#interface port-channel 1
AS2(config-if)#switchport mode trunk
```

步骤三：查看设备捆绑端口

```
AS1#show etherchannel summary
Flags:   D - down         P - in port-channel
         I - stand-alone  s - suspended
         H - Hot-standby (LACP only)
         R - Layer3       S - Layer2
         U - in use       f - failed to allocate aggregator
         u - unsuitable for bundling
         w - waiting to be aggregated
         d - default port


Number of channel-groups in use: 1
Number of aggregators:           1

Group   Port-channel        Protocol        Ports
```

```
----+-----------+---------+------------------------------------
1       Po1(SU)                -           Gig0/1(P) Gig0/2(P)
```

从上面显示的状态可知,端口为聚合端口。生成树状态如下。

```
AS1#show spanning-tree
VLAN0001
  Spanning tree enabled protocol ieee
  Root ID    Priority      32769
             Address       0001.C992.0871
             This bridge is the root
             Hello Time  2 sec   Max Age 20 sec   Forward Delay 15 sec

  Bridge ID  Priority      2769   (priority 32768 sys-id-ext 1)
             Address       0001.C992.0871
             Hello Time 2 sec   Max Age 20 sec   Forward Delay 15 sec
             Aging Time 20

Interface        Role Sts    Cost       Prio.Nbr Type
---------------- ---- ---  ---------   -------- --------------------------------
Fa0/1            Desg FWD    19         128.1   P2p
Po1              Desg FWD    3          128.27  Shr
```

我们还可以配置动态二层链路聚合,配置端口所属通道组和工作模式如下。

(1)在交换机 AS1 上配置动态二层链路聚合

```
AS1(config)#interface range GigabitEthernet 0/1-2
AS1(config-if-range)# channel-group 1 mode active
AS1(config-if-range)#interface port-channel 1
AS1(config-if)# switchport mode trunk
```

(2)在交换机 AS2 上配置动态二层链路聚合

```
AS2(config)#interface range GigabitEthernet 0/1-2
AS2(config-if-range)# channel-group 1 mode passive
AS2(config-if-range)#interface port-channel 1
AS2(config-if)# switchport mode trunk
```

静态链路聚合不用任何协议,可强制形成链路聚合状态;动态链路聚合需要采用 PAgP 或 LACP,通过协商形成链路聚合状态。

1.2.3 场景七：配置三层链路聚合

交换机 CS1 和交换机 CS2 为核心交换机，交换机的 Gi1/1/1 和 Gi1/1/2 端口为光端口，交换机的 Gi1/0/22~24 三个端口为 RJ-45 端口。为了增加带宽和进行冗余备份，采用交换机的两条光纤链路为主链路进行三层链路聚合，采用交换机的三条双绞线链路为备份链路进行三层链路聚合；要求光纤链路采用静态链路聚合，双绞线链路采用动态链路聚合。三层链路聚合拓扑如图 1-10 所示。

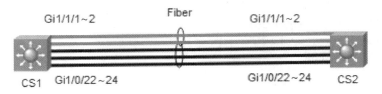

图 1-10 三层链路聚合拓扑

具体配置步骤如下所述。

步骤一：配置三层静态链路聚合

（1）配置 CS1 的光纤链路聚合

```
CS1(config)#interface range GigabitEthernet 1/1/1-2
CS1(config-if-range)#channel-group 10 mode on
CS1(config-if-range)#exit
CS1(config)#interface port-channel 10
CS1(config-if)#no switchport
CS1(config-if)#ip address 10.0.0.1 255.255.255.252
CS1(config-if)#no shut
```

（2）配置 CS2 的光纤链路聚合

```
CS2(config)#interface range GigabitEthernet 1/1/1-2
CS2(config-if-range)#channel-group 10 mode on
CS2(config-if-range)#exit
CS2(config)#interface port-channel 10
CS2(config-if)#no switchport
CS2(config-if)#ip address 10.0.0.2 255.255.255.252
CS2(config-if)#no shut
```

步骤二：配置三层动态链路聚合

（1）配置 CS1 的双绞线链路聚合

```
CS1(config)#interface range GigabitEthernet 1/0/22-24
CS1(config-if-range)#channel-protocol lacp
CS1(config-if-range)#channel-group 20 mode active
CS1(config-if-range)#exit
CS1(config)#interface port-channel 20
CS1(config-if)#no switchport
CS1(config-if)#ip add 10.0.1.1 255.255.255.252
CS1(config-if)#no shut
```

（2）配置 CS2 的双绞线链路聚合

```
CS2(config)#interface range GigabitEthernet 1/0/22-24
CS2(config-if-range)#channel-protocol lacp
CS2(config-if-range)#channel-group 30 mode passive
CS2(config-if-range)#exit
CS2(config)#interface port-channel 30
CS2(config-if)#no switchport
CS2(config-if)#ip add 10.0.1.2 255.255.255.252
CS2(config-if)#no shut
```

步骤三：查看端口配置

（1）查看 CS1 链路聚合端口状态

```
CS1#show ip interface brief | include Port-channel
Port-channel10         10.0.0.1         YES manual up          up
Port-channel20         10.0.1.1         YES manual up          up
```

（2）查看 CS2 链路聚合端口状态

```
CS2#show ip interface brief | include Port-channel
Port-channel10         10.0.0.2         YES manual up          up
Port-channel30         10.0.1.2         YES manual up          up
```

（3）查看 CS1 链路聚合状态

```
CS1#show etherchannel summary
Flags:  D - down         P - in port-channel
```

```
            I - stand-alone s - suspended
            H - Hot-standby (LACP only)
            R - Layer3      S - Layer2
            U - in use      f - failed to allocate aggregator
            u - unsuitable for bundling
            w - waiting to be aggregated
            d - default port
Number of channel-groups in use: 2
Number of aggregators:           2

Group   Port-channel            Protocol        Ports
------+-------------+-----------+------------------------------------------

10      Po10(RU)                -               Gig1/1/1(P) Gig1/1/2(P)
20      Po20(RU)                LACP            Gig1/0/22(P) Gig1/0/23(P) Gig1/0/24(P)
```

1.2.4 场景八：分析链路聚合故障

某网络工程师为了提高级联链路的带宽，采用链路聚合技术将 4 条物理链路进行捆绑，但发现并没有达到预期效果，进一步检查后发现聚合链路有问题。如图 1-11 所示，分析链路聚合故障。

图 1-11　分析链路聚合故障

情境一：待绑定的物理端口工作模式不匹配，导致聚合链路中部分物理链路不能正常工作，由下述可知，在 S1 的聚合链路中，Gi1/0/3 和 Gi1/0/4 两个物理端口处于非正常状态。

```
S1#show etherchannel summary
Flags: D - down P - in port-channel
       I - stand-alone s - suspended
       H - Hot-standby (LACP only)
       R - Layer3 S - Layer2
       U - in use f - failed to allocate aggregator
       u - unsuitable for bundling
       w - waiting to be aggregated
       d - default port
```

Number of channel-groups in use: 1
Number of aggregators: 1

Group Port-channel Protocol   Ports
------+-------------+-----------+---

1   Po1(SU)   -   Gig1/0/1(P) Gig1/0/2(P) **Gi1/0/3(s) Gi1/0/4(s)**
S1#

在交换机上查看端口工作模式，发现交换机 S1 的端口 Gi1/0/3 被手工配置为 trunk 模式，而 Gi1/0/1 端口被配置为 dynamic auto 模式。

```
S1#show interfaces g1/0/3 switchport
Name: Gig1/0/3
Switchport: Enabled
Administrative Mode: trunk
Operational Mode: down
Administrative Trunking Encapsulation: dot1q
Operational Trunking Encapsulation: dot1q
Negotiation of Trunking: On
Access Mode VLAN: 1 (default)
Trunking Native Mode VLAN: 1 (default)
……………………………………
```

```
S1#show interfaces g1/0/1 switchport
Name: Gig1/0/1
Switchport: Enabled
Administrative Mode: dynamic auto
Operational Mode: static access
Administrative Trunking Encapsulation:
Operational Trunking Encapsulation: native
Negotiation of Trunking: On
Access Mode VLAN: 1 (default)
Trunking Native Mode VLAN: 1 (default)
……………………………………
```

通过命令 **show running-config** 查看运行的配置命令。

```
S1#show running-config
Building configuration...
```

......
interface GigabitEthernet1/0/1
channel-group 1 mode active
!
interface GigabitEthernet1/0/2
channel-group 1 mode active
!
interface GigabitEthernet1/0/3
switchport trunk encapsulation dot1q
switchport mode trunk
channel-group 1 mode active
!
interface GigabitEthernet1/0/4
channel-group 1 mode active
!

故障解决方法：将 S1 的 Gi1/0/3 端口设置为与绑定组中的其余端口相同的属性即可。

S1#**configure terminal**
S1(config)#**interface g1/0/3**
S1(config-if)#**no switchport trunk encapsulation dot1q**
S1(config-if)#**switchport mode dynamic auto**
S1(config-if)#**end**
S1#**show etherchannel summary**
Flags: D - down   P - in port-channel
       I - stand-alone   s - suspended
       H - Hot-standby (LACP only)
       R - Layer3   S - Layer2
       U - in use   f - failed to allocate aggregator
       u - unsuitable for bundling
       w - waiting to be aggregated
       d - default port
Number of channel-groups in use: 1
Number of aggregators: 1

Group Port-channel Protocol   Ports
------+-------------+-----------+---

1 0   Po1(SU)   Gig1/0/1(P) Gig1/0/2(P) **Gig1/0/3(P) Gig1/0/4(P)**

S1#

注意：trunk 端口的封装协议必须采用默认的协商方式配置，不能采用手工配置，否则会影响绑定后的逻辑链路的形成。如果将 Gi1/0/3 端口恢复为默认属性，逻辑链路仍不能形成，可以将设备重启一下。

情境二：待绑定的物理端口上允许不同 VLAN 的数据通过，导致聚合链路中部分物理链路不能正常工作。由下述可知，交换机 S1 的聚合链路中 Gi1/0/2 端口不能正常工作，交换机 S2 的聚合链路中只有 Gi1/0/1 端口能正常工作，其余端口均处于非正常状态。

```
S1#show etherchannel summary
Flags: D - down        P - in port-channel
       I - stand-alone s - suspended
       H - Hot-standby (LACP only)
       R - Layer3      S - Layer2
       U - in use      f - failed to allocate aggregator
       u - unsuitable for bundling
       w - waiting to be aggregated
       d - default port


Number of channel-groups in use: 1
Number of aggregators:           1

Group  Port-channel  Protocol    Ports
------+-------------+-----------+-------------------------------------------
1      Po1(SU)       -           Gig1/0/1(P) Gig1/0/2(s) Gig1/0/3(P) Gig1/0/4(P)
S1#
```

```
S2#show etherchannel summary
Flags: D - down        P - in port-channel
       I - stand-alone s - suspended
       H - Hot-standby (LACP only)
       R - Layer3      S - Layer2
       U - in use      f - failed to allocate aggregator
       u - unsuitable for bundling
       w - waiting to be aggregated
       d - default port
```

```
             Number of channel-groups in use: 1
             Number of aggregators: 1
             Group Port-channel Protocol Ports
------+-------------+-----------+---------------------------------------
10    Po1(SU) - Gi1/0/1(P) Gi1/0/2(s) Gig1/0/3(s) Gi1/0/4(s)
S2#
```

通过在交换机 S1 上显示运行配置文件中的配置命令，可以看到 Gi1/0/2 端口被设置为不允许 VLAN 30 的数据通过。

```
S1#show running-config
Building configuration...
...............................
interface Port-channel1
switchport trunk encapsulation dot1q
switchport mode trunk
!
interface GigabitEthernet1/0/1
switchport trunk encapsulation dot1q
switchport mode trunk
channel-group 1 mode on
!
interface GigabitEthernet1/0/2
switchport trunk allowed vlan 1-29,31-1005
switchport trunk encapsulation dot1q
switchport mode trunk
channel-group 1 mode on
!
interface GigabitEthernet1/0/3
switchport trunk encapsulation dot1q
switchport mode trunk
channel-group 1 mode on
!
interface GigabitEthernet1/0/4
switchport trunk encapsulation dot1q
switchport mode trunk
channel-group 1 mode on
!
```

也可以通过以下命令进一步验证，交换机 S1 的 Gi1/0/2 端口上不允许 VLAN 30 的数据通过。

```
S1#show interfaces g1/0/2 switchport
Name: Gig1/0/2
Switchport: Enabled
Administrative Mode: dynamic auto
Operational Mode: static access
Administrative Trunking Encapsulation: negotiated
Operational Trunking Encapsulation: native
Negotiation of Trunking: On
Access Mode VLAN: 1 (default)
Trunking Native Mode VLAN: 1 (default)
Voice VLAN: none
Administrative private-vlan host-association: none
Administrative private-vlan mapping: none
Administrative private-vlan trunk native VLAN: none
Administrative private-vlan trunk encapsulation: dot1q
Administrative private-vlan trunk normal VLANs: none
Administrative private-vlan trunk private VLANs: none
Operational private-vlan: none
Trunking VLANs Enabled: 1-29,31-1005
Pruning VLANs Enabled: 2-1001
Capture Mode Disabled
Capture VLANs Allowed: ALL
Protected: false
```

交换机 S1 的其余端口允许所有 VLAN 的数据通过。

```
S1#show interfaces g1/0/1 switchport
Name: Gig1/0/1
Switchport: Enabled
Administrative Mode: dynamic auto
Operational Mode: down
Administrative Trunking Encapsulation: negotiated
Operational Trunking Encapsulation: dot1q
Negotiation of Trunking: On
Access Mode VLAN: 1 (default)
Trunking Native Mode VLAN: 1 (default)
Voice VLAN: none
Administrative private-vlan host-association: none
Administrative private-vlan mapping: none
Administrative private-vlan trunk native VLAN: none
Administrative private-vlan trunk encapsulation: dot1q
```

Administrative private-vlan trunk normal VLANs: none
Administrative private-vlan trunk private VLANs: none
Operational private-vlan: none
Trunking VLANs Enabled: All
Pruning VLANs Enabled: 2-1001
Capture Mode Disabled
Capture VLANs Allowed: ALL
Protected: false
Appliance trust: none

故障解决方法：将交换机 S1 的 Gi1/0/2 端口设置为允许所有 VLAN 的数据通过，与绑定组中的其余端口属性一致。

S1#**configure terminal**
S1(config)#**interface g1/0/2**
S1(config-if)#**switchport trunk allowed vlan all**
S1(config-if) #**end**
S1#**show etherchannel summary**
Flags: D - down P - in port-channel
　　　I - stand-alone s - suspended
　　　H - Hot-standby (LACP only)
　　　R - Layer3 S - Layer2
　　　U - in use f - failed to allocate aggregator
　　　u - unsuitable for bundling
　　　w - waiting to be aggregated
　　　d - default port

Number of channel-groups in use: 1
Number of aggregators: 1
Group Port-channel Protocol Ports
------+-------------+-----------+---
1　　Po1(SU) - Gig1/0/1(P) Gig1/0/2(P) Gig1/0/3(P) Gig1/0/4(P)
S1#

注意：如果待绑定的端口存在双工模式、速率等方面的不匹配，均可能导致捆绑失败。

1.3 学习网关冗余技术

1.3.1 认识网关冗余

网关冗余技术是大型网络中不可缺少的技术。通常，可通过配置 HSRP（Hot Standby Router

Protocol，热备份路由器协议）和 VRRP（Virtual Router Redundancy Protocol，虚拟路由器冗余协议）来实现网关冗余，由此出现两种实现网关冗余的方案：HSRP 方案和 VRRP 方案。HSRP 和 VRRP 都是解决局域网中配置静态网关出现单点失效现象的路由协议。HSRP 和 VRRP 比较如表 1-5 所示。

表 1-5 HSRP 与 VRRP 比较

HSRP（热备份路由器协议）	VRRP（虚拟路由器冗余协议）
Cisco 私有协议	由 IETF 提出
最多支持 255 个组	最多支持 255 个组
1 台活跃路由器、1 台备份路由器、若干台候选路由器	1 台活跃路由器、若干台候选路由器
虚拟 IP 地址与真实 IP 地址不能相同	虚拟 IP 地址与真实 IP 地址可以相同
使用 224.0.0.2 发送 Hello 数据包	使用 224.0.0.18 发送 Hello 数据包
默认计时器：Hello 时间为 3 s，保持时间为 10 s	默认计时器：默认小于 HSRP
可以追踪端口或对象	只能追踪对象
支持认证	支持认证

网关冗余工作原理：由一台活跃的路由器充当网关路由器，另一台或多台其他路由器则处于备用状态，一旦活跃路由器失效，备份路由器可以马上接替其角色。由两台或者多台路由器共享一个 IP 地址和 MAC 地址，共同维护一个虚拟路由器。

1.3.2 剖析网关冗余

1. HSRP 消息类型

① Hello 消息：Hello 消息被用于通知其他路由器发送消息的路由器的 HSRP 优先级和状态信息，配置 HSRP 的路由器默认每 3 s 发送 1 个 Hello 消息。

② 政变（coup）消息：当 1 台备份路由器变为 1 台活跃路由器时会发送 1 个 coup 消息。

③ 辞职（resign）消息：当活跃路由器要宕机或者当有优先级更高的路由器发送 Hello 消息时，活跃路由器发送一个 resign 消息。

2. HSRP 路由器的状态

表 1-6 给出了 HSRP 中路由器的 5 种状态。

表 1-6 HSRP 方案中路由器的 5 种状态

初始状态（Initial）	未运行 HSRP 时的状态
学习状态（Learn）	等待活跃路由器发送 Hello 消息
监听状态（Listen）	路由器知道虚拟 IP 地址，但还不是活跃路由器和备份路由器，同时监听 Hello 消息
发言状态（Speak）	周期性发送 Hello 消息，并积极参与活跃路由器和备份路由器选举

备用状态（Standby）	该路由器是成为下一个活跃路由器的候选设备，周期性发送 Hello 消息
活跃状态（Active）	路由器负责转发发送到 HSRP 组的虚拟 MAC 地址的数据包，并周期性发送 Hello 消息

1.3.3　场景九：配置单组 HSRP

某公司为了防止链路出故障无法访问外网，将路由器 R1 和 R2 分别与外网连接，如图 1-12 所示，该图展示了单组 HSRP 网络拓扑。平时只需要主链路（R1—ISP）转发数据，另一条为备份链路（R2—ISP）。当路由器 R1 失效时，处于备用状态的路由器 R2 必须快速切换为活跃状态，接替路由器 R1 的角色。

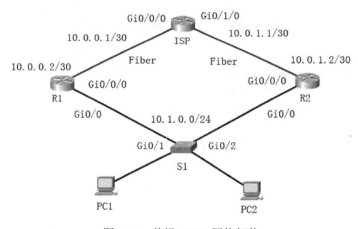

图 1-12　单组 HSRP 网络拓扑

HSRP 链路聚合的语法如下：

interface *interface*	//进入接口模式
standby *group-number* ip *ip-address*	//配置 HSRP 组及虚拟路由器 IP 地址
standby *group-number* preempt	//配置 HSRP 抢占模式
standby *group-number* priority *priority-value*	//配置 HSRP 组优先级

具体配置步骤如下所述。

步骤一：配置三层端口 IP 地址

（1）配置 ISP 端口地址

```
ISP(config)#interface GigabitEthernet 0/0/0
ISP(config-if)#ip address 10.0.0.1 255.255.255.252
ISP(config-if)#no shutdown
ISP(config-if)#exit
```

```
ISP(config)#interface GigabitEthernet 0/1/0
ISP(config-if)#ip address 10.0.1.1 255.255.255.252
ISP(config-if)#no shutdown
```

(2) 配置 R1 端口地址

```
R1(config)#interface GigabitEthernet 0/0/0
R1(config-if)#ip address 10.0.0.2 255.255.255.252
R1(config-if)#no shutdown
R1(config-if)#exit
R1(config)#interface GigabitEthernet 0/0
R1(config-if)#ip address 10.1.0.252 255.255.255.0
R1(config-if)#no shutdown
```

(3) 配置 R2 端口地址

```
R2(config)#interface GigabitEthernet 0/0/0
R2(config-if)#ip address 10.0.1.2 255.255.255.252
R2(config-if)#no shutdown
R2(config-if)#exit
R2(config)#interface GigabitEthernet 0/0
R2(config-if)#ip address 10.1.0.253 255.255.255.0
R2(config-if)#no shutdown
```

步骤二：在端口上配置 HSRP

(1) 在 R1 端口上配置 HSRP

```
R1(config)#interface GigabitEthernet 0/0
R1(config-if)#standby 10 ip 10.1.0.254
R1(config-if)#standby 10 preempt
R1(config-if)#standby 10 priority 110
```

(2) 在 R2 端口上配置 HSRP

```
R2(config)#interface GigabitEthernet 0/0
R2(config-if)#standby 10 ip 10.1.0.254
R2(config-if)#standby 10 preempt
```

步骤三：查看 HSRP 的配置

(1) 在 R1 上查看 HSRP 的配置

```
R1#show standby brief
```

P indicates configured to preempt.

Interface	Grp	Pri	P	State	Active	Standby	Virtual IP
Gig0/0	10	110	P	Active	local	10.1.0.253	10.1.0.254

（2）在 R2 上查看 HSRP 的配置

R2#**show standby brief**

P indicates configured to preempt.

Interface	Grp	Pri	P	State	Active	Standby	Virtual IP	
Gig0/0	10	100	P	Standby	10.1.0.252	local	10.1.0.254	a

1.3.4　场景十：配置多组 HSRP

如图 1-13 所示，局域网内有多个 VLAN，分别为 VLAN 10、VLAN 20、VLAN 30 和 VLAN 40，原网络中的所有数据都从交换机 CS1 转发到外网，交换机 CS2 为备份设备，虽然每天也在运行，但是不参与网络中的数据转发。为了防止设备资源的浪费，现需要在交换机 CS1 和 CS2 上进行负载均衡配置，使 VLAN 10 和 VLAN 20 的数据流通过交换机 CS1 转发，当交换机 CS1 失效时数据流量被切换到交换机 CS2；使 VLAN 30 和 VLAN 40 的数据流通过交换机 CS2 转发，当交换机 CS2 失效时，数据流量被切换到交换机 CS1。这样可以防止设备闲置造成的资源浪费，提高网络性能。图 1-13 展示了多组 HSRP 网络拓扑，表 1-7 展示了 HSRP 参数表。

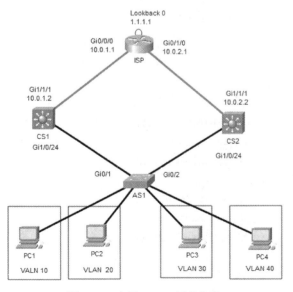

图 1-13　多组 HSRP 网络拓扑

表 1-7　HSRP 参数表

VLAN	端口分配	HSRP 组号	HSRP 虚拟 IP 地址
VLAN 10	Fa0/1~5	10	10.1.1.254/24
VLAN 20	Fa0/6~10	20	10.1.2.254/24
VLAN 30	Fa0/11~15	30	10.1.3.254/24
VLAN 40	Fa0/16~20	40	10.1.4.254/24

具体配置步骤如下。

步骤一：创建 VLAN 并划分端口

（1）在 CS1 上创建 VLAN

```
CS1(config)#vlan 10
CS1(config-vlan)#vlan 20
CS1(config-vlan)#vlan 30
CS1(config-vlan)#vlan 40
```

（2）在 CS2 上创建 VLAN

```
CS2(config)#vlan 10
CS2(config-vlan)#vlan 20
CS2(config-vlan)#vlan 30
CS2(config-vlan)#vlan 40
```

（3）在 AS1 上创建 VLAN 并划分端口

```
AS1(config)#vlan 10
AS1(config-vlan)#vlan 20
AS1(config-vlan)#vlan 30
AS1(config-vlan)#vlan 40
AS1(config)#interface range FastEthernet 0/1-5
AS1(config-if-range)#switchport mode access
AS1(config-if-range)#switchport access vlan 10
AS1(config-if-range)#exit
AS1(config)#interface range FastEthernet 0/6-10
AS1(config-if-range)#switchport mode access
AS1(config-if-range)#switchport access vlan 20
AS1(config-if-range)#exit
AS1(config)#interface range FastEthernet 0/11-15
AS1(config-if-range)#switchport mode access
```

```
AS1(config-if-range)#switchport access vlan 30
AS1(config-if-range)#exit
AS1(config)#interface range FastEthernet 0/16-20
AS1(config-if-range)#switchport mode access
AS1(config-if-range)#switchport access vlan 40
```

步骤二：配置 trunk 端口

（1）在 CS1 上配置 trunk 端口

```
CS1(config)#interface GigabitEthernet 1/0/24
CS1(config-if)#switchport trunk encapsulation dot1q
CS1(config-if)#switchport mode trunk
```

（2）在 CS2 上配置 trunk 端口

```
CS2(config)#interface GigabitEthernet 1/0/24
CS2(config-if)#switchport trunk encapsulation dot1q
CS2(config-if)#switchport mode trunk
```

（3）在 AS1 上配置 trunk 端口

```
AS1(config)#interface range GigabitEthernet 0/1-2
AS1(config-if-range)#switchport mode trunk
```

步骤三：配置端口地址

（1）配置 ISP 端口地址

```
ISP(config)#interface GigabitEthernet 0/0/0
ISP(config-if)#ip address 10.0.1.1 255.255.255.252
ISP(config-if)#no shutdown
ISP(config-if)#exit
ISP(config)#interface GigabitEthernet 0/1/0
ISP(config-if)#ip address 10.0.2.1 255.255.255.252
ISP(config-if)#no shutdown
ISP(config)#interface loopback 0
ISP(config-if)#ip address 1.1.1.1 255.255.255.255
```

（2）配置 CS1 端口地址

```
CS1(config)#interface GigabitEthernet 1/1/1
CS1(config-if)#no switchport
```

```
CS1(config-if)#ip address 10.0.1.2 255.255.255.252
CS1(config-if)#exit
CS1(config)#interface vlan 10
CS1(config-if)#ip address 10.1.1.252 255.255.255.0
CS1(config-if)#exit
CS1(config)#interface vlan 20
CS1(config-if)#ip address 10.1.2.252 255.255.255.0
CS1(config-if)#exit
CS1(config)#interface vlan 30
CS1(config-if)#ip address 10.1.3.252 255.255.255.0
CS1(config-if)#exit
CS1(config)#interface vlan 40
CS1(config-if)#ip address 10.1.4.252 255.255.255.0
CS1(config-if)#exit
```

（3）配置 CS2 端口地址

```
CS2(config)#interface GigabitEthernet 1/1/1
CS2(config-if)#no switchport
CS2(config-if)#ip address 10.0.2.2 255.255.255.252
CS2(config-if)#no shutdown
CS2(config-if)#exit
CS2(config)#interface vlan 10
CS2(config-if)#ip address 10.1.1.253 255.255.255.0
CS2(config-if)#exit
CS2(config)#interface vlan 20
CS2(config-if)#ip address 10.1.2.253 255.255.255.0
CS2(config-if)#exit
CS2(config)#interface vlan 30
CS2(config-if)#ip address 10.1.3.253 255.255.255.0
CS2(config-if)#exit
CS2(config)#interface vlan 40
CS2(config-if)#ip address 10.1.4.253 255.255.255.0
CS2(config-if)#exit
```

步骤四：开启交换机路由转发功能并配置静态路由

（1）在 CS1 上配置静态路由

```
CS1(config)#ip routing
```

```
CS1(config)#ip route 0.0.0.0 0.0.0.0 10.0.1.1
```

(2) 在 CS2 上配置静态路由

```
CS2(config)#ip routing
CS2(config)#ip route 0.0.0.0 0.0.0.0 10.0.2.1
```

步骤五：配置 HSRP

(1) 在 CS1 上配置 HSRP

```
CS1(config)#interface vlan 10
CS1(config-if)#standby 10 ip 10.1.1.254
CS1(config-if)#standby 10 preempt
CS1(config-if)#standby 10 priority 120
CS1(config-if)#exit
CS1(config)#interface vlan 20
CS1(config-if)#standby 20 ip 10.1.2.254
CS1(config-if)#standby 20 preempt
CS1(config-if)#standby 20 priority 120
CS1(config-if)#exit
CS1(config)#interface vlan 30
CS1(config-if)#standby 30 ip 10.1.3.254
CS1(config-if)#standby 30 preempt
CS1(config-if)#exit
CS1(config)#interface vlan 40
CS1(config-if)#standby 40 ip 10.1.4.254
CS1(config-if)#standby 40 preempt
```

(2) 在 CS2 上配置 HSRP

```
CS2(config)#interface vlan 10
CS2(config-if)#standby 10 ip 10.1.1.254
CS2(config-if)#standby 10 preempt
CS2(config-if)#exit
CS2(config)#interface vlan 20
CS2(config-if)#standby 20 ip 10.1.2.254
CS2(config-if)#standby 20 preempt
CS2(config-if)#exit
CS2(config)#interface vlan 30
CS2(config-if)#standby 30 ip 10.1.3.254
```

```
CS2(config-if)#standby 30 preempt
CS2(config-if)#standby 30 priority 120
CS2(config-if)#exit
CS2(config)#interface vlan 40
CS2(config-if)#standby 40 ip 10.1.4.254
CS2(config-if)#standby 40 preempt
CS2(config-if)#standby 40 priority 120
```

步骤六：查看 HSRP 配置

（1）在 CS1 上查看 HSRP 配置

```
CS1#show standby brief
                     P indicates configured to preempt.
                      |
Interface   Grp   Pri P State   Active       Standby       Virtual IP
Vl10        10    120 P Active  local        10.1.1.253    10.1.1.254
Vl20        20    120 P Active  local        10.1.2.253    10.1.2.254
Vl30        30    100 P Standby 10.1.3.253   local         10.1.3.254
Vl40        40    100 P Standby 10.1.4.253   local         10.1.4.254
```

（2）在 CS2 上查看 HSRP 配置

```
CS2#show standby brief
                     P indicates configured to preempt.
                      |
Interface   Grp   Pri P State   Active       Standby       Virtual IP
Vl10        10    100 P Standby 10.1.1.252   local         10.1.1.254
Vl20        20    100 P Standby 10.1.2.252   local         10.1.2.254
Vl30        30    120 P Active  local        10.1.3.252    10.1.3.254
Vl40        40    120 P Active  local        10.1.4.252    10.1.4.254
```

由以上输出可以看出，CS1 是 VLAN 10 和 VLAN 20 的活跃网关，CS2 是 VLAN 30 和 VLAN 40 的活跃网关。

1.4 挑战练习

挑战要求：

① 为确保网络的高可用性，采用链路聚合技术，实现上行链路带宽的扩展。具体做法是分别采用二层、三层链路聚合，以提高整个网络的运行效率。

② 为保证网络二层链路运行的高可靠性，采用 MSTP 技术并实现负载均衡；要求所有用户均能快速接入网络，并且能防止在接终端设备的端口上接入交换设备。正常情况下 LAN 1 的流量经过 DS1 转发，LAN 2 的流量经过 DS2 转发，当某一台汇聚层设备发生故障时，流量会自动切换到另外一台设备上。

③ 为保证网络三层链路运行的高可靠性，配置网关冗余技术，确保虚拟网关中的活跃端口随着二层 STP 的主根自动切换。

挑战练习实验拓扑如图 1-14 所示，请读者独立完成该挑战练习。

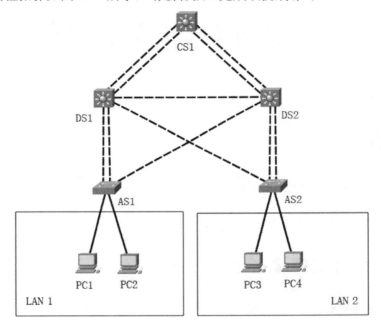

图 1-14　挑战练习实验拓扑

1.5　本章小结

本章内容到此结束。本章主要内容包括 STP、链路聚合和网关冗余。在本章的学习中，我们通过 10 个有趣的应用场景和 1 个挑战练习来加深对交换冗余技术的理解。在现实网络中，交换冗余技术的应用十分广泛，该技术的使用为网络二层链路提供了更高的可靠性和稳定性，因此希望读者熟练掌握本章内容。

第2章 >>>

学习 LAN 安全技术

本章要点：

- 学习端口安全技术
- 学习 IEEE 802.1x 认证技术
- 学习 DHCP 侦听技术
- 学习防止 ARP 攻击技术
- 学习 STP 优化技术
- 挑战练习
- 本章小结

本章通过介绍各种 LAN 安全技术，帮助读者通过所学技术增强 LAN 的安全性。其中，2.1 节介绍端口安全技术，通过设置 MAC 地址与交换机端口绑定来限制终端用户的访问；2.2 节介绍 IEEE 802.1x 认证技术，通过认证限制未授权用户的接入；2.3 节介绍 DHCP 侦听技术，通过设置交换机端口为可信或不可信端口来拦截 DHCP 攻击；2.4 节介绍防止 ARP 攻击技术；2.5 节通过介绍 STP 优化技术提升网络性能。本章设计了 6 个应用场景和 1 个挑战练习，可让读者深入理解所学内容，轻松掌握相关知识，灵活应用所学技术。

2.1 学习端口安全技术

2.1.1 认识端口安全

1．端口安全原理

可通过 MAC 地址表记录连接到交换机端口的以太网卡的 MAC 地址，并且只允许某个/些 MAC 地址通过本端口通信，当其他 MAC 地址发送的数据包通过此端口时，端口安全特性会阻止该数据包。

2．端口安全作用

使用端口安全特性可以防止未经允许的设备访问网络，既可增强网络安全性，又可防止因 MAC 地址泛洪造成 MAC 地址表被填满。

3．端口安全类型

① 静态端口安全：使用命令在端口上手动绑定 MAC 地址，所绑定的 MAC 地址被静态存储在地址表中，不会随着端口的重新启动而被移除。

② 动态端口安全：通过动态方式获取 MAC 地址并将其绑定到相关端口，所绑定的 MAC 地址在端口重新启动时将被移除。

③ 粘滞端口安全：通过动态方式获取 MAC 地址，所获取 MAC 地址将被静态存储到地址表中，同时被添加到运行配置文件中。用此方式绑定的 MAC 地址也不会随着端口重新启动而被移除，当需要将大量 MAC 地址静态绑定到某端口时，通常采用粘滞端口安全。

4．端口安全违规动作

当交换机端口收到"非法"流量时会执行相应的操作，即端口安全违规动作。端口安全违规动作及其含义如表 2-1 所示。

表 2-1 端口安全违规动作及其含义

违规动作	含义
保护（protect）	丢弃未允许的 MAC 地址流量，不创建日志消息
限制（restrict）	丢弃未允许的 MAC 地址流量，创建日志消息并发送 SNMP Trap 消息
关闭（shutdown）	将端口设置为 err-disabled 状态，创建日志消息并发送 SNMP Trap 消息。需要手动恢复或者使用 err-disable recovery 特性重新开启该端口

2.1.2 场景一：配置静态绑定 MAC 地址的端口安全

如图 2-1 所示，在交换机 S1 上配置静态绑定 MAC 地址的端口安全功能，使 PC1 可以通过交换机 S1 的 Fa0/1 端口访问 LAN 中其他网络，而当除 PC1 之外的其他设备通过交换机 S1 的 Fa0/1 端口访问 LAN 时，使该端口关闭（shutdown）。

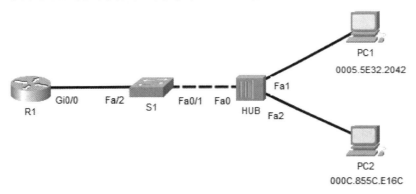

图 2-1 静态绑定 MAC 地址的端口安全

端口安全配置语法如下：

```
switch port mode access                                              //必须指定交换机端口为 access 模式
switchport port-security                                             //启用端口安全
switchport port-security maximum max-number                          //配置端口安全最大数量
switchport port-security mac-address XXXX.XXXX.XXXX                  //配置静态绑定 MAC 地址
switchport port-security violation { protect | restrict | shutdown } //配置端口安全违规动作
```

在交换机 S1 上配置端口安全功能，静态绑定 PC1 的 MAC 地址并将端口安全违规动作设置为 shutdown。

```
S1(config)#interface FastEthernet 0/1
S1(config-if)#switchport mode access
S1(config-if)#switchport port-security
S1(config-if)#switchport port-security maximum 1                     //默认配置，可省略
S1(config-if)#switchport port-security mac-address 0005.5E32.2042
```

S1(config-if)#**switchport port-security violation shutdown**   //默认配置，可省略

分别使用 PC1 和 PC2 访问路由器 R1 并查看结果。

（1）通过 PC1 访问路由器 R1

```
C:\>ping 10.1.1.1

Pinging 10.1.1.1 with 32 bytes of data:

Reply from 10.1.1.1: bytes=32 time=1ms TTL=255
Reply from 10.1.1.1: bytes=32 time<1ms TTL=255
Reply from 10.1.1.1: bytes=32 time<1ms TTL=255
Reply from 10.1.1.1: bytes=32 time<1ms TTL=255

Ping statistics for 10.1.1.1:
    Packets: Sent = 4, Received = 4, Lost = 0 (0% loss),
Approximate round trip times in milli-seconds:
Minimum = 0ms, Maximum = 1ms, Average = 0ms
```

（2）通过 PC2 访问路由器 R1

```
C:\>ping 10.1.1.1

Pinging 10.1.1.1 with 32 bytes of data:

Request timed out.
Request timed out.
Request timed out.
Request timed out.

Ping statistics for 10.1.1.1:
    Packets: Sent = 4, Received = 0, Lost = 4 (100% loss),
```

交换机端口状态调试如图 2-2 所示，由该图可知，当 PC2 访问换机 S1 的 Fa0/1 端口时，该端口被关闭。因为 PC2 的 MAC 地址不是端口安全绑定的 MAC 地址。

```
%LINK-5-CHANGED: Interface FastEthernet0/1, changed state to
administratively down

%LINEPROTO-5-UPDOWN: Line protocol on Interface FastEthernet0/1,
changed state to down
```

图 2-2　交换机端口状态调试

（3）查看交换机 S1 的 Fa0/1 端口状态

```
S1#show port-security interface fastEthernet 0/1
Port Security                  : Enabled
Port Status                    : Secure-shutdown
Violation Mode                 : Shutdown
Aging Time                     : 0 mins
Aging Type                     : Absolute
SecureStatic Address Aging     : Disabled
Maximum MAC Addresses          : 1
Total MAC Addresses            : 1
Configured MAC Addresses       : 1
Sticky MAC Addresses           : 0
Last Source Address:Vlan       : 000C.855C.E16C:1
Security Violation Count       : 10
```

由上面的状态可以看到，Fa0/1 端口已经被关闭，这时需要管理员手动开启端口。

```
S1(config)#interface FastEthernet 0/1
S1(config-if)#shutdown
S1(config-if)#no shutdown
```

2.1.3 场景二：配置粘滞绑定 MAC 地址端口安全

如图 2-3 所示，在交换机 S1 上配置粘滞绑定 MAC 地址端口安全功能，端口最大允许连接的 MAC 地址数量为 2，当该端口连接两台设备后，其他设备的流量不能通过交换机 S1 的 Fa0/1 端口。如果其他设备违规访问，将使其端口处于保护（protect）状态。

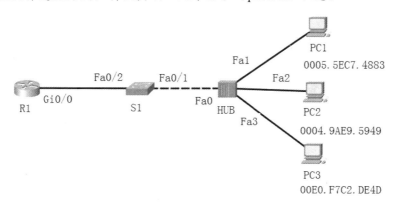

图 2-3　粘滞绑定 MAC 地址端口安全

在交换机 S1 上设置粘滞端口安全功能，使端口最大允许的 MAC 地址数量为 2，如果违规将使相应端口处于保护（protect）状态。

```
S1(config)#interface FastEthernet 0/1
S1(config-if)#switchport mode access
S1(config-if)#switchport port-security
S1(config-if)#switchport port-security mac-address sticky
S1(config-if)#switchport port-security maximum 2
S1(config-if)#switchport port-security violation protect
```

使用 PC1、PC2 和 PC3 分别访问路由器 R1 并查看结果。

（1）通过 PC1 访问路由器 R1

```
C:\>ping 10.1.1.1

Pinging 10.1.1.1 with 32 bytes of data:

Reply from 10.1.1.1: bytes=32 time=1ms TTL=255
Reply from 10.1.1.1: bytes=32 time<1ms TTL=255
Reply from 10.1.1.1: bytes=32 time<1ms TTL=255
Reply from 10.1.1.1: bytes=32 time<1ms TTL=255

Ping statistics for 10.1.1.1:
    Packets: Sent = 4, Received = 4, Lost = 0 (0% loss),
Approximate round trip times in milli-seconds:
Minimum = 0ms, Maximum = 1ms, Average = 0ms
```

（2）通过 PC2 访问路由器 R1

```
C:\>ping 10.1.1.1

Pinging 10.1.1.1 with 32 bytes of data:

Reply from 10.1.1.1: bytes=32 time<1ms TTL=255
Reply from 10.1.1.1: bytes=32 time<1ms TTL=255
Reply from 10.1.1.1: bytes=32 time<1ms TTL=255
Reply from 10.1.1.1: bytes=32 time<1ms TTL=255

Ping statistics for 10.1.1.1:
    Packets: Sent = 4, Received = 4, Lost = 0 (0% loss),
```

Approximate round trip times in milli-seconds:
Minimum = 0ms, Maximum = 0ms, Average = 0ms

（3）通过 PC3 访问路由器 R1

```
C:\>ping 10.1.1.1

Pinging 10.1.1.1 with 32 bytes of data:

Request timed out.
Request timed out.
Request timed out.
Request timed out.

Ping statistics for 10.1.1.1:
    Packets: Sent = 4, Received = 0, Lost = 4 (100% loss),
```

（4）检查安全 MAC 地址

```
S1#show port-security address
            Secure Mac Address Table
-------------------------------------------------------------
Vlan   Mac Address      Type           Ports           Remaining Age
                                                       (mins)
----   -----------      ----           -----           ------------
 1     0004.9AE9.5949   SecureSticky   FastEthernet0/1    -
 1     0005.5EC7.4883   SecureSticky   FastEthernet0/1    -
-------------------------------------------------------------
Total Addresses in System (excluding one mac per port)     : 1
Max Addresses limit in System (excluding one mac per port) : 1024
```

由上可知，交换机 S1 与 PC1 和 PC2 连接后，交换机 S1 的 Fa0/1 端口绑定的 MAC 地址数量已为 2，与设定的 maximum 值相同，所以 PC3 的数据从端口 Fa0/1 通过时会被丢弃。

我们再看 S1 交换机的配置文件，从中可以看到，端口安全功能已经将 PC1 和 PC2 两台设备的 MAC 地址写入到运行配置文件中。这时对运行配置文件进行保存，重启设备后粘滞的 MAC 地址不会丢失。

```
S1#show run | begin interface FastEthernet0/1
interface FastEthernet0/1
 switchport mode access
 switchport port-security
```

```
switchport port-security maximum 2
switchport port-security mac-address sticky
switchport port-security violation protect
switchport port-security mac-address sticky 0004.9AE9.5949
switchport port-security mac-address sticky 0005.5EC7.4883
```

2.2 学习 IEEE 802.1x 认证技术

2.2.1 认识 IEEE 802.1x 认证

IEEE 802.1x 协议标准可以限制未经授权的用户通过接入端口访问 LAN，在获得 LAN 提供的各种业务前，需要对接入交换机的用户或设备进行认证，认证通过后相关用户或设备方可通过端口访问交换机。

2.2.2 场景三：配置 IEEE 802.1x 认证

在图 2-4 中，AAA Server 负责提供 IEEE 802.1x 认证服务，IP 地址为 10.1.1.2/24；TFTP Server 负责提供交换机 S1 版本升级所需的 ZOS 映像，IP 地址为 10.1.1.4/24。先将交换机 S1 的 IOS 版本进行升级（注意：交换机必须进行版本升级，否则不支持 IEEE 802.1x 认证），然后在交换机 S1 上完成 IEEE 802.1x 相关配置，使 PC1 必须通过认证后才可以访问内网和外网。配置 IEEE 802.1x 认证拓扑如图 2-4 所示。

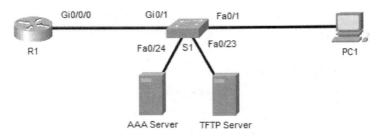

图 2-4 配置 IEEE 802.1x 认证拓扑

步骤一：升级交换机 IOS 版本

（1）升级交换机版本

```
S1(config)#vlan 10
S1(config-vlan)#name AUTH
S1(config-vlan)#exit
```

```
S1(config)#interface range f 0/1-24
S1(config-if-range)#switchport mode access
S1(config-if-range)#switchport access vlan 10
S1(config-if-range)#inter vlan 10
S1(config-if)#ip add 10.1.1.10 255.255.255.0
S1(config-if)#no shutdown
S1(config-if)#end
S1#copy tftp: flash:
Address or name of remote host []? 10.1.1.4
Source filename []? c2960-lanbasek9-mz.150-2.SE4.bin
Destination filename [c2960-lanbasek9-mz.150-2.SE4.bin]?

Accessing tftp://10.1.1.4/c2960-lanbasek9-mz.150-2.SE4.bin.....
Loadingc2960-lanbasek9-mz.150-2.SE4.bin from 10.1.1.4:
!!!!!!!!!!!!!!!!!!!!!!!!!!!!!!!!!!!!!!!!!!!!!!!!!!!!!!!!!!!!!!!!!!!!!!!!!!!!!!!!!!!!!!!
[OK - 4670455 bytes]

4670455 bytes copied in 11.073 secs (33910 bytes/sec)
S1#delete flash:c2960-lanbase-mz.122-25.FX.bin
Delete filename [c2960-lanbase-mz.122-25.FX.bin]?c2960-lanbase-mz.122-25.FX.bin
Delete flash:/c2960-lanbase-mz.122-25.FX.bin? [confirm]y
S1#reload
```

（2）查看交换机版本

在交换机 S1 上查看交换机版本，检查是否升级成功。

```
S1#show version | begin Switch Ports
Switch Ports Model                SW Version        SW Image
------ ----- -----                ----------        ----------
  * 1    26  WS-C2960-24TT-L      15.0(2)SE4        C2960-LANBASEK9-M


Configuration register is 0xF
```

步骤二：在服务器上配置 AAA 认证

如图 2-5 和图 2-6 所示，在 AAA Server 上配置 Radius 服务并开启认证功能。

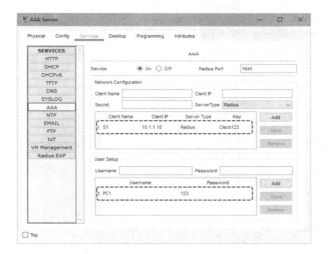

图 2-5 在 AAA Server 上配置 Radius 服务

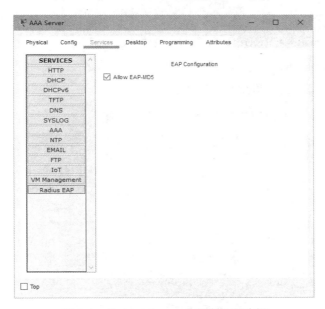

图 2-6 在 AAA Server 上开启认证功能

步骤三：在交换机 S1 上配置 IEEE 802.1x 认证功能

```
S1(config)#aaa new-model
S1(config)#radius-server host 10.1.1.2 auth-port 1645 key Cisco123
S1(config)#aaa authentication dot1x default group radius
S1(config)#dot1x system-auth-control
S1(config)#interface range FastEthernet 0/1-15
```

S1(config-if-range)#**switchport mode access**
S1(config-if-range)#**authentication port-control auto**
S1(config-if-range)#**dot1x pae authenticator**

步骤四：验证 IEEE 802.1x 功能

（1）在认证前通过 PC1 访问外网

C:\>**ping 10.1.1.2**

Pinging 10.1.1.2 with 32 bytes of data:

Request timed out.
Request timed out.
Request timed out.
Request timed out.

Ping statistics for 10.1.1.2:
    Packets: Sent = 4, Received = 0, Lost = 4 (100% loss),

结果：认证前访问不成功。

（2）在 PC1 上配置 IEEE 802.1x 认证功能

如图 2-7 所示，在 PC1 上配置 IEEE 802.1x 认证功能。

图 2-7　在 PC1 上配置 IEEE 802.1x 认证功能

（3）PC1 通过认证后访问外网

```
C:\>ping 10.1.1.2

Pinging 10.1.1.2 with 32 bytes of data:

Reply from 10.1.1.2: bytes=32 time=3ms TTL=128
Reply from 10.1.1.2: bytes=32 time<1ms TTL=128
Reply from 10.1.1.2: bytes=32 time=1ms TTL=128
Reply from 10.1.1.2: bytes=32 time<1ms TTL=128

Ping statistics for 10.1.1.2:
    Packets: Sent = 4, Received = 4, Lost = 0 (0% loss),
Approximate round trip times in milli-seconds:
    Minimum = 0ms, Maximum = 3ms, Average = 1ms
```

结果：认证通过后访问成功。

2.3 学习 DHCP 侦听技术

2.3.1 认识 DHCP 攻击

1．DHCP 欺骗

攻击者在局域网内伪装成一台 DHCP 服务器，对合法客户的 DHCP 请求做出响应，为客户提供虚假的 IP 地址和网关。

2．DHCP 耗竭

攻击者在局域网内伪装成 DHCP 客户端，无限制申请 IP 地址，使 DHCP 服务器的 IP 地址枯竭。通常两种攻击伴随发生，一般先进行 DHCP 耗竭攻击，再进行 DCHP 欺骗攻击。

3．DHCP 侦听

当交换机开启 DHCP 侦听后，交换机会对 DHCP 报文进行侦听，并从接收到的 DHCP Request 或 DHCP Ack 报文中提取地址信息。当某个物理端口被配置为信任端口或不信任端口时，信任端口可以正常接收和发送 DHCP Offer 报文，不信任端口会丢弃 DHCP Offer 报文。

2.3.2 场景四：配置 DHCP 侦听

在图 2-8 中，DHCP Server 是 LAN 内合法 DHCP 服务器，为 LAN 内终端设备提供 DHCP 服务。现有一台 PC 为非法客户端，无限制地申请 IP 地址，使地址枯竭，从而使 DHCP Server 无法提供 DHCP 服务，而由非法 DHCP Server 为 LAN 内客户端提供 DHCP 服务。现要求在交换机 S1 上配置 DHCP Snooping 功能，使每台终端设备发送的 DHCP 请求报文数量得到控制，并通过使用信任端口使非法 DHCP Server 无法为 LAN 内客户端提供非法服务。配置 DHCP 侦听拓扑如图 2-8 所示。

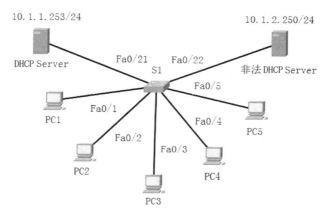

图 2-8　配置 DHCP 侦听拓扑

DHCP Snooping 语法规则如下：

ip dhcp snooping	//开启 DHCP Snooping 功能
ip dhcp snooping trust	//在端口模式下，将端口配置为信任端口
ip dhcp snooping limit rate *num*	//在端口模式下，设置单位时间内接收报文数量

在交换机 S1 上配置 DHCP Snooping 功能，将 DHCP Server 所在端口配置为信任端口（默认全部为非信任端口）。为防止非法 DHCP 客户端使地址枯竭，从而使 DHCP Server 无法提供 DHCP 服务，限制端口报文接收速率。

```
S1(config)#ip dhcp snooping
S1(config)#interface FastEthernet 0/21
S1(config-if)#ip dhcp snooping trust
S1(config-if)#exit
S1(config)#interface range FastEthernet 0/1-5
S1(config-if-range)#ip dhcp snooping limit rate 15
```

在交换机上开启 DHCP 侦听功能，对 DHCP 报文进行窥探。如果网络中存在 DHCP 攻击行为，将会消耗交换机性能，应该通过命令在端口上配置允许接收的 DHCP 报文数量。

查看 DHCP 侦听的配置信息：

```
S1#show ip dhcp snooping
Switch DHCP snooping is enabled
DHCP snooping is configured on following VLANs:
none
Insertion of option 82 is enabled
Option 82 on untrusted port is not allowed
Verification of hwaddr field is enabled
Interface              Trusted     Rate limit (pps)
-----------------      -------     ----------------
FastEthernet0/1        no          15
FastEthernet0/2        no          15
FastEthernet0/3        no          15
FastEthernet0/4        no          15
FastEthernet0/5        no          15
FastEthernet0/21       yes         unlimited
```

由于非法 DHCP Server 为非信任端口，不能发送 DHCP Offer 数据包，因此不会为 LAN 内设备提供 DHCP 服务。

接下来我们将合法 DHCP Server 服务器移除，查看非法 DHCP Server 是否会提供 DHCP 服务。非法 DHCP Server 服务配置如图 2-9 所示，从其配置信息可以看到，非法 DHCP Server 正在正常提供 DHCP 服务。

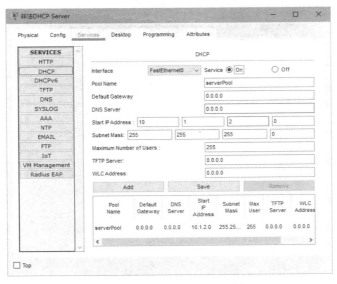

图 2-9 非法 DHCP Server 服务配置

如图 2-10 所示，查看 PC3 获取地址情况。从图中可知，终端 PC3 无法正常获取到 IP 地址。

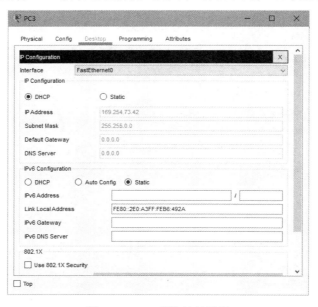

图 2-10　PC3 获取地址情况

接下来，将合法 DHCP Server 通过信任端口接入到网络中，再次查看 PC3 地址获取情况。如图 2-11 所示，PC3 成功获取 IP 地址。

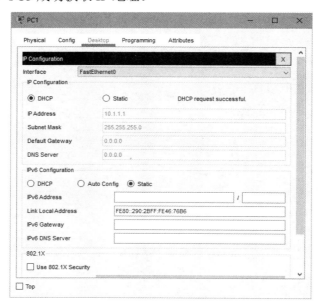

图 2-11　PC3 成功获取 IP 地址

2.4 学习防止 ARP 攻击技术

2.4.1 认识 DAI 的原理及作用

1. DAI 定义

DAI：Dynamic ARP Inspection，动态 ARP 检测。通过进行 DAI，可在交换机上提供 IP 地址和 MAC 地址绑定功能并动态建立绑定关系。

2. ARP 欺骗原理

当用户发送 ARP 广播请求消息时，攻击主机会伪装成网关欺骗用户，然后将其 MAC 地址写入 ARP 响应消息回发给用户；同时攻击主机也伪装成发送数据的用户欺骗网关，欺骗网关把 ARP 响应消息发送给自己，这样用户所有的数据都将经过攻击主机之后才会转发到网关。

3. DHCP Snooping+DAI 原理

当开启 DHCP Snooping 功能后，用户会维护一个 IP 地址与 MAC 地址的绑定信息表，DAI 可以利用 DHCP Snooping 的绑定信息表来检查信任端口的 ARP 请求和响应，确保数据是用户发送的，来自虚拟 MAC 地址的则会被丢弃数据包。

2.4.2 场景五：配置 DHCP Snooping 及 DAI

在图 2-12 中，PC2 为攻击者，想要截获所有 PC1 发送的数据包，使 PC1 发送的数据包经过 PC2 转发给路由器 R1。我们任务是：通过采用 DHCP Snooping+DAI 解决方案来解决 ARP 欺骗问题并找到正确的网关。为了完成任务，请在交换机 S1 上进行相关配置。DHCP Snooping 及 DAI 应用拓扑如图 2-12 所示。

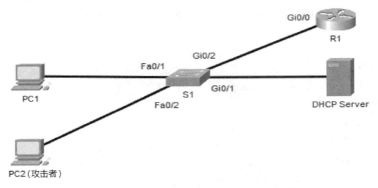

图 2-12　DHCP Snooping 及 DAI 应用拓扑

在交换机上开启 DHCP Snooping 及 DAI 功能，使 PC1 所在 VLAN 中启用 DHCP 监听和 ARP 检测功能。

```
S1(config)#vlan 10
S1(config-vlan)#name dhcp
S1(config)#interface range FastEthernet 0/1-20
S1(config-if-range)#switchport mode access
S1(config-if-range)#switchport access vlan 10
S1(config)#ip dhcp snooping
S1(config)#interface GigabitEthernet 0/1
S1(config-if-range)#ip dhcp snooping trust
S1(config-if-range)#ip arp inspection trust
S1(config-if-range)#exit
S1(config)#ip dhcp snooping vlan 10
S1(config)#ip arp inspection vlan 10
S1(config)#ip arp inspection validate src-mac dst-mac ip
```

查看 DAI 信息：

```
S1#show ip arp inspection

Source Mac Validation        : Enabled
Destination Mac Validation   : Enabled
IP Address Validation        : Enabled
```

Vlan	Configuration	Operation	ACL Match	Static ACL
10	Enabled	Inactive		

Vlan	ACL Logging	DHCP Logging	Probe Logging
10	Deny	Deny	Off

Vlan	Forwarded	Dropped	DHCP Drops	ACL Drops
10	166	5	5	0

Vlan	DHCP Permits	ACL Permits	Probe Permits	Source MAC Failures
10	104	0	0	0

Vlan	Dest MAC Failures	IP Validation Failures	Invalid Protocol Data
10	0	0	0

2.5 学习 STP 优化技术

2.5.1 认识 STP 的优化参数

有关 STP 端口状态以及转换过程的知识，我们在第 1 章中已经学习过，详见表 1-2 和图 1-2。

① PortFast：快速端口，一般在交换机连接终端设备的端口上配置。设置了该选项的端口可绕过 Listening 和 Learning 状态直接进入 Forwarding 状态，极大地减少了设备接入的时间。

② BPDU Guard：BPDU 保护，如果配置了该选项的端口收到 BPDU 报文，则端口会立即切换到 err-disable 状态。该选项常与 PortFast 选项一起使用，用于交换机连接终端的端口。

③ Root Guard：根保护，是强制保护 STP 根的措施，即防止新加入网络的交换机（有更低网桥 ID）成为网络中的根桥，从而影响交换网络的稳定性，此选项需要在指定端口上配置。

④ UplinkFast：思科专有特性，能够在上连链路失效时提供快速收敛能力，即减少 STP 收敛时间，适用于接入层交换机。当交换机配置了该选项后，如果 Forwarding 状态的端口失效，Blocking 状态的端口将会跳过 Listening 和 Learning 姿态，直接进入 Forwarding 状态。

⑤ BackboneFast：是对 UplinkFast 的一种补充，UplinkFast 用于检测直连链路失效，BackboneFast 用于检测间接链路失效。当启用 BackboneFast 的交换机检测到间接链路失效时，会使 Blocking 状态的端口进入 Listening 状态，减少 20 s 的老化时间，要求所有交换机都启用。

注意：UplinkFast 和 BackboneFast 特性是针对 IEEE 802.1d 传统 STP 进行的优化，而 RSTP（IEEE 802.1w）和 MSTP（IEEE 802.1s）已内置集成了类似 UplinkFast 和 BackboneFast 的功能，因此当采用 RSTP 和 MSTP 时，无须也无法激活这两种特性。

2.5.2 场景六：优化 STP 的网络性能

如图 2-13 所示，将 3 台交换机互连形成一个物理环路，在该交换网络中，交换机 S1 为接入层交换机，我们希望降低交换机间的 STP 收敛过程对接入终端设备的影响，使终端设备能尽快接入网络；如果再接入新交换机，不允许参与 STP 根选举，以免影响网络的稳定性；对于交换机连接终端设备的端口，要求开启 BPDU Guard 功能以阻止私自接入交换机。请配置 STP 的优化参数，提升网络性能。如果 S2 为根交换机，则 S1 的 Fa0/17 端口将被选为根端口，我们希望当其 Fa0/17 端口出现故障时，能快速切换到 Fa0/16 端口继续转发数据，以缩短网络中断时间。可使用 STP 的 UplinkFast 特性来降低当链路故障时的网络丢包率。

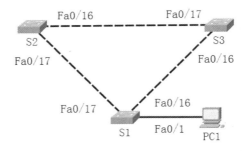

图 2-13 优化 STP 的网络性能拓扑

PortFast、BPDU Guard、Root Guard、UplinkFast 和 BackboneFast 的配置语法如下：

spanning-tree portfast	//在端口模式下，开启 PortFast 功能
spanning-tree portfast bpdugurad default	//在全局配置模式下，开启 BPDU Guard 功能
spanning-tree guard root	//在端口模式下，开启 STP 根保护功能
spanning-tree uplinkfast	//在全局配置模式下，开启 UplinkFast 功能
spanning-tree backbonefast	//在全局配置模式下，开启 BackboneFast 功能

具体配置步骤如下。

步骤一：配置 PortFast 和 BPDU Guard 功能

（1）在 S1 交换机上配置 PortFast 功能

```
S1(config)#interface FastEthernet 0/1
S1(config-if)#shutdown
S1(config-if)#no shutdown
```

在启动 PortFast 功能之前，端口 Fa0/1 的 STP 状态处于 Listening 状态，如下所示。

```
S1#show spanning-tree
<output omitted>
Interface        Role Sts Cost      Prio.Nbr Type
---------------- ---- --- --------- -------- --------
Fa0/16           Desg FWD 19        128.16   P2p
Fa0/17           Desg FWD 19        128.17   P2p
Fa0/1            Desg LSN 19        128.1    P2p  //LSN 状态，即 Listening 状态
```

根据 IEEE 802.1d 规则，Listening 状态持续 15 s，随后进入 Learning 状态，再持续 15 s，读者可自行计时观察。综上，该端口从启动开始共计 30 s 内不能转发用户数据，即图 2-13 中的 PC1 在这 30 s 内无法连接网络。

在交换机 S1 的端口 Fa0/1 启动 PortFast 功能，观察 STP 状态的变化：

```
S1(config)#interface FastEthernet 0/1
S1(config-if)#spanning-tree portfast
S1(config-if)#shutdown
S1(config-if)#no shutdown
```

在端口 Fa0/1 启动 PortFast 后,关闭并开启该端口,其立即进入 Forwarding 状态,如下所示。

```
S1#show spanning-tree
<output omitted>
Interface        Role Sts Cost      Prio.Nbr Type
---------------- ---- --- --------- --------
Fa0/16           Desg FWD 19        128.16   P2p
Fa0/17           Desg FWD 19        128.17   P2p
Fa0/1            Desg FWD 19        128.1    P2p    //FWD 状态,即 Forwarding 状态
```

与不启用 PortFast 功能相比,端口提前 30 s 进入转发状态,所连接的 PC1 可立即接入网络。

(2)在交换机 S1 上配置 BPDU Guard(BPDU 保护)功能

思科交换机允许在两个级别上配置启用 BPDU Guard 功能。

① 在全局配置模式下,针对配置了 PortFast 功能的所有端口启用 BPDU Guard,命令如下:

```
S1(config)#spanning-tree portfast bpduguard default
```

注意,该命令只针对启动了 PortFast 功能的端口,当这些端口上收到 BPDU 数据包时立即进入 err-disable 状态,端口将被关闭。

② 在端口配置模式下,启用 BPDU Guard,命令如下:

```
S1(config-if)#spanning-tree bpduguard enable
```

该命令仅对所选端口生效,无论端口是否启用 PortFast 功能,只要在端口上收到 BPDU 报文,该端口立即进入 err-disable 状态并被关闭。

(3)配置 PortFast 和 BPDU Guard 功能后在交换机 S1 上接入终端设备

```
S1#show spanning-tree
VLAN0001
  Spanning tree enabled protocol ieee
  Root ID    Priority     32769
             Address      04fe.7f30.6200
             This bridge is the root
             Hello Time   2 sec   Max Age 20 sec   Forward Delay 15 sec
  Bridge ID  Priority     32769    (priority 32768 sys-id-ext 1)
             Address      04fe.7f30.6200
```

```
                    Hello Time 2 sec  Max Age 20 sec   Forward Delay 15 sec
                    Aging Time  300 sec
Interface           Role Sts         Cost           Prio.Nbr  Type
---------------- ---- --- --------- --------------------------------
Fa0/1               Desg FWD         19             128.3     P2p
```

（4）配置 PortFast 和 BPDU Guard 功能后在交换机 S1 的 Fa0/1 端口上接入交换机

```
*Mar  1 01:46:16.533: %SPANTREE-2-BLOCK_BPDUGUARD: Received BPDU on port Fa/0/1 with
BPDU Guard enabled. Disabling port.     //在端口 Fa0/1 上收到 BPDU 报文，关闭该端口
*Mar  1 01:46:16.533: %PM-4-ERR_DISABLE: bpduguard error detected on Fa0/1, putting Fa0/1 in
err-disable state          //在端口 Fa0/1 上检测到 BPDU 报文，将该端口设置为 err-disable 状态
S1#show interfaces fastEthernet 0/1
FastEthernet0/1 is down, line protocol is down (err-disabled)
```

当端口模式配置 spanning-tree bpduguard enable 后，无论是否在端口上启用 PortFast 功能，只要端口收到 BPDU 报文都会显示上面的提示信息，端口立即进入 err-disable 状态并被关闭。

步骤二：配置 UplinkFast 功能

（1）观察交换机开启 UplinkFast 功能之前上连端口失效时的生成树收敛情况

在图 2-13 所示拓扑中，S2 为根交换机，S1 的 Fa0/17 为根端口，Fa0/16 是 NDP（Non-Designated Port，非指定端口），处于 Blocking 状态。下面我们将 S1 的根端口 Fa0/17 设置为失效状态，观察 STP 状态变化。

```
S1(config)#interface FastEthernet 0/17
S1(config-if)#shutdown     //将交换机 S1 的根端口关闭模拟链路失效
```

根端口 Fa0/17 失效后，根据 IEEE 802.1d 规则，端口 Fa0/16 立即进入 Listening 状态，如下所示。

```
S1#show spanning-tree
<output omitted>
Interface          Role Sts Cost    Prio.Nbr Type
---------------- ---- --- --------- --------------------------------
Fa0/16             Root LIS 19      128.16   P2p   //端口 Fa0/16 进入 Listening 状态
Fa0/1              Desg FWD 19      128.1    P2p
```

S1 的端口 Fa0/16 在 Listening 状态持续 15 s，随后进入 Learning 状态，如下所示。

```
S1#show spanning-tree    //历时 15 s 后执行
<output omitted>
Interface          Role Sts Cost    Prio.Nbr Type
---------------- ---- --- --------- --------------------------------
```

Fa0/16	Root LRN 19		128.16	P2p	//端口 Fa0/16 进入 Learning 状态
Fa0/1	Desg FWD 19		128.1	P2p	

S1 的端口 Fa0/16 在 Learning 状态再持续 15 s，然后进入 Forwarding 状态，如下所示。

```
S1#show spanning-tree       //再历时 15 s 后再次执行
<output omitted>
Interface        Role Sts Cost      Prio.Nbr Type
---------------- ---- --- --------- --------
```

Fa0/16	Root FWD 19		128.16	P2p	//端口 Fa0/16 进入 Forwarding 状态
Fa0/1	Desg FWD 19		128.1	P2p	

此时端口 Fa0/16 成为新的根端口并开始转发用户数据。综上，当 S1 根端口失效后，要等待 30 s，新的根端口才开始转发数据。如何加快 STP 收敛，缩短这个时间，思科专属 UplinkFast 功能可解决这个问题。

（2）在接入层交换机 S1 上开启 UplinkFast 功能

注意：目前 PT 8.0 版本不支持 UplinkFast 功能，以下配置输出内容是真实交换机的输出结果。

将交换机 S1 的根端口关闭模拟链路失效：

```
S1(config)#spanning-tree uplinkfast
S1(config)#interface fastEthernet 0/17
S1(config-if)#shutdown
*Jul 10 16:17:01.335: %SPANTREE_FAST-7-PORT_FWD_UPLINK: VLAN0001 FastEthernet0/16
moved to Forwarding (UplinkFast).
```

以上输出信息表明，在启动 UplinkFast 功能后，一旦 S1 当前根端口失效，非指定端口（NDP）Fa0/16 将立即转变为 Forwarding 状态，开始转发用户数据。

（3）查看交换机 S1 的生成树状态

```
S1#show spanning-tree
VLAN0001
  Spanning tree enabled protocol ieee
  Root ID    Priority    32768
             Address     04fe.7f30.7c80
             Cost        3019
             Port        19 (FastEthernet1/0/17)
             Hello Time  2 sec   Max Age 20 sec   Forward Delay 15 sec

  Bridge ID  Priority    49153    (priority 49152 sys-id-ext 1)
             Address     04fe.7f30.6200
             Hello Time  2 sec   Max Age 20 sec   Forward Delay 15 sec
```

```
                    Aging Time      15   sec
         uplinkfast enabled
         Interface          Role   Sts     Cost        Prio.Nbr    Type
         ---------------    ----   ---     --------    --------    -----
         Fa0/16             Root   FWD     3019        128.19      P2p
```

完成以上操作后，我们发现 S1 与 S3 相连的端口 Fa0/16 的 Cost 值增加了 3000，同时 S1 优先级被自动设置为 49152。因为 S1 的端口 Fa0/17 失效后，端口 Fa0/16 成为新的转发端口。

启动 UplinkFast 后，端口的 Cost 值增加了 3000，并且交换机优先级被置为 49152（默认为 32768），以避免此接入层交换机变为转接交换机。所谓转接交换机是指该交换机转发的数据流量的目的地不是本交换机所连接的终端设备，即流量从某个交换机级联端口进入，从另外的交换机级联端口送出。在局域网设计中，通常将核心层和分布层交换机作为转接交换机，而 Uplinkfast 功能针对的是接入层交换机。因此，增加端口 Cost 数值和交换机优先级，可避免在启用 Uplinkfast 功用后，接入层交换机变为转接交换机。

综上所述，UplinkFast 功能针对接入层交换机，并且需要在 NDP 所在的接入层交换机上启用，当交换机发现某直连端口发生故障时，UplinkFast 功能会立即启用备用端口。

注意：一个良好的 STP 设计方案，接入层交换机通常不作为根交换机。

步骤三：配置 Root Guard 功能

在图 2-13 的拓扑中 S2 为根交换机，为了保护其根交换机的角色，避免其他交换机（包括潜在的新接入网络的交换机）成为新的根交换机，在 S2 的两个指定端口 Fa0/16 和 Fa0/17 上启用 Root Guard 功能。在下面的配置过程中，更改 S1 的优先级使其小于 S2 的优先级，然后观察 S2 交换机 Fa0/16 和 Fa0/17 的端口状态。

（1）在交换机 S2 上开启 Root Guard 功能

```
S2(config)#interface range f0/16-17
S2(config-if-range)#spanning-tree guard root
```

（2）设置交换机 S1 的优先级尝试使其成为根交换机

```
S1(config)#spanning-tree vlan 1 priority 4096
```

（3）查看交换机 S2 的状态

目前，PT 8.0 版本不能完全模拟真实交换机设备上的根端口保护状态，下列内容是真实交换机上输出的结果。

```
%SPANTREE-2-ROOTGUARDBLOCK: Port 0/17 tried to become non-designated in VLAN 1.
Moved to root-inconsistent state
%SPANTREE-2-ROOTGUARDBLOCK: Port 0/16 tried to become non-designated in VLAN 1.
Moved to root-inconsistent state
```

```
S2#show spanning-tree
VLAN0001
  Spanning tree enabled protocol ieee
  Root ID    Priority      32769
             Address       04fe.7f30.6200
             This bridge is the root
             Hello Time   2 sec  Max Age 20 sec  Forward Delay 15 sec

  Bridge ID  Priority      32769   (priority 32768 sys-id-ext 1)
             Address       04fe.7f30.6200
             Hello Time   2 sec  Max Age 20 sec  Forward Delay 15 sec
             Aging Time    15 sec

Interface           Role Sts       Cost         Prio.Nbr     Type
---- --- --------   ----------------------------------------------------------
Fa0/16              Desg BKN*      19           128.16       P2p *ROOT_Inc
Fa0/17              Desg BKN*      19           128.17       P2p *ROOT_Inc
```

上面显示的端口状态 BKN 表示线路处于断路状态，这是开启 Root Guard 功能后收到含有低于自身优先级 BPDU 报文时所采取的保护性措施。

综上，开启 Root Guard 功能后，如果收到更低优先级的 BPDU 报文，则会造成网络中断，必须采取恢复措施，例如，调整非根交换机的优先级，以恢复网络连接。

步骤四：配置 BackboneFast 功能

UplinkFast 功能用于在接入层交换机发生直连线路故障时立即启用备用端口，加快 STP 收敛。而 BackboneFast 功能用于间接线路故障的探测，仍是加快 STP 收敛速度。

在图 2-13 所示的拓扑中，S2 为根交换机，S1 的端口 Fa0/17 为根端口，端口 Fa0/16 是 NDP，处于 Blocking 状态。当 S3 的根端口 Fa0/17 失效后，因为 S1 的端口 Fa0/16 处于 Blocking 状态，不会向 S3 发送 BPDU 报文，所以 S3 无法收到当前根交换机的信息。

根据 IEEE 802.1d 规则，S3 将主动从其仅有工作端口 Fa0/16 发送 BPDU 报文，声称自己是根交换机。S1 的 NDP（端口 Fa0/16）收到这些 BPDU 报文后会忽略它们，因为此时 S1 能够从其端口 Fa0/17 收到当前根交换机 S2 发出的 BPDU 报文，其优先级优于 S3。S1 等待 20 s（默认的 max age 超时时间）后，其端口 Fa0/16 端口进入 Listening 状态，此时 S1 向 S3 转发它从 S2 收到的 BPDU 报文，随后，当 S3 知晓当前根交换机 S2 工作正常后，停止主动发送 BPDU 报文。S1 的端口 Fa0/16 的 Listening 状态持续 15 s 后进入 Learning 状态，再持续 15 s，共 30 s，然后进入 Forwarding 状态，此时端口 Fa0/16 成为指定端口并开始转发用户数据。随后 S3 的端口 Fa0/16 成为新的根端口，STP 重新收敛。整个收敛过程持续 50 s（S1 端口 Fa0/16 首先等待 20 s，然后经过两个时间长度为 15 s 的过渡状态，在此期间交换机不转发用户数据，因此终端设备处于断网状态。

我们希望缩短收敛过程，加快收敛，BackboneFast 功能是解决该问题的思科专属功能。它的改进之处在于，在上述过程中，当 S1 的端口 Fa0/16 收到 S3 发出的 BPDU 报文后不再等待时长为 20 s 的 max age 超时时间，而是主动从它自己的根端口发送 RLQ (Root Link Query) BPDU 报文，询问当前根交换机是否正常。S2 收到 RLQ BPDU 报文后将回复 S1，然后 S1 端口 Fa0/16 进入 Listening 状态并向 S3 发送 BPDU，通知当前根交换机工作正常。随后 S1 的端口 Fa0/16 的 Listening 状态持续 15 s 后进入 Learning 状态，再持续 15 s，共 30 s，然后进入 Forwarding 状态。因此，BackboneFast 功能使整个收敛时间缩短 20 s，由原来 50 s 变为 30 s。

（1）观察交换机在开启 BackboneFast 功能之前非直连端口失效时 STP 收敛情况

```
S3(config)#interface FastEthernet 0/17
S3(config-if)#shutdown
```

将交换机 S3 的根端口 Fa0/17 关闭模拟链路失效，我们在交换机 S1 上观察生成树的变化。

```
S1# debug spanning-tree events
Spanning Tree event debugging is on
*Jul 11 17:15:33.729: STP: VLAN0001 heard root 32769-fa16.3e7c.0e00 on Fa0/16
<output omitted>
*Jul 11 17:15:49.739: STP: VLAN0001 heard root 32769-fa16.3e7c.0e00 on Fa0/16
*Jul 11 17:15:51.565: STP: VLAN0001 Fa0/16 ->listening
*Jul 11 17:15:51.741: STP: VLAN0001 heard root 32769-fa16.3e7c.0e00 on Fa0/16
*Jul 11 17:16:06.566: STP: VLAN0001 Fa0/16 ->learning
*Jul 11 17:16:21.567: STP: VLAN0001 Fa0/16 ->forwarding
```

从以上输出时钟信息可知，整个收敛时间历时约 50 s（时钟信息存在误差，非严格 50 s）。

（2）在所有交换机上开启 BackboneFast 功能

注意：目前 PT 8.0 版本不支持 BackboneFast 功能，以下交换机配置输出内容基于真实交换机。

```
S1(config)#spanning-tree backbonefast
S2(config)#spanning-tree backbonefast
S3(config)#spanning-tree backbonefast
S3(config)#interface FastEthernet 0/17
S3(config-if)#no shutdown        //恢复交换机 S3 根端口 Fa0/17 的工作状态
```

（3）查看交换机启用 BackboneFast 功能后生成树收敛时间的变化

```
S3(config)#interface FastEthernet 0/17
S3(config-if)#shutdown
```

将交换机 S3 根端口 Fa0/17 关闭模拟链路失效，我们在交换机 S1 上观察生成树的变化。

```
S1# debug spanning-tree events
*Jul 11  17:30:40.720: STP: VLAN0001 heard root 32769-fa16.3e7c. 0e00 on Fa0/16
*Jul 11  17:30:40.723: STP: VLAN0001 Fa0/16 -> listening
*Jul 11  17:30:55.724: STP: VLAN0001 Fa0/16 ->learning
```

```
*Jul 11    17:31:10.726: STP: VLAN0001 Fa0/16 ->forwarding
```

从以上输出时钟信息可知，整个收敛时间历时 30 s，与启用 BackboneFast 功能前相比缩短了 20 s）。

2.6　挑战练习

挑战要求：

① 配置 STP，实现负载均衡。正常情况下 LAN 1 的流量经过 DS1 转发，LAN 2 的流量经过 DS2 转发，当某一台汇聚层设备发生故障时，流量会自动切换到另外一台设备上。所有用户均能快速接入网络，并能防止在连接终端设备的端口上接入交换设备。

② 配置网关冗余，确保虚拟网关中的活跃端口随着二层 STP 的主根自动切换。

③ 完成 DHCP 侦听的配置，保证所有 PC 只能从合法的 DHCP 处获得 IP 地址，防止合法 DHCP 服务器的地址被耗竭；对所有主机进行合适配置，防止 ARP 欺骗的发生。

④ 对于 LAN 1 的用户，通过 IEEE 802.1x 进行认证，只允许该网络中的合法用户访问网络；对于 LAN 2 的用户，通过端口安全访问网络，确保每个端口只允许某个特定的合法用户接入网络，拒绝非法用户的流量并且在日志中进行记录。

挑战练习实验拓扑如图 2-14 所示，请读者独立完成该挑战练习。

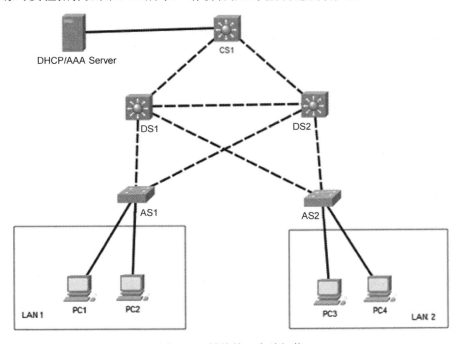

图 2-14　挑战练习实验拓扑

2.7 本章小结

本章内容到此结束。本章主要内容包括端口安全、IEEE 802.1x 认证、DHCP 侦听、ARP 攻击和 STP 优化。在网络技术的发展过程中，人们渐渐发现了 LAN 的一些弊端，例如，ARP 欺骗和 DHCP 欺骗等，本章所介绍的技术正是为了克服此类弊端。在本章的学习中，我们通过 6 个应用场景和 1 个挑战练习来帮助大家掌握安全技术的应用，借助 LAN 安全技术，我们可以提高 LAN 的安全性，提升网络性能。

第3章 >>>

学习基础 OSPFv2

本章要点：

- 学习 OSPFv2
- 配置基础 OSPFv2
- 检验 OSPFv2 配置
- 干预 OSPFv2 运行
- 挑战练习
- 本章小结

本章介绍了 OSPFv2 相关基础知识。其中，3.1 节介绍 OSPFv2 基本概念，包括 OSPFv2 的特征、数据库、原理、数据包类型、运行状态以及路由器类型；3.2 节介绍基础 OSPFv2 配置，包括路由器 ID、单区域 OSPFv2 以及多区域 OSPFv2 的配置；3.3 节介绍检验 OSPFv2 配置相关内容，包括检验 OSPFv2 配置，查看 OSPFv2 单区域和多区域的三张表以及 LSA 类型、验证 OSPFv2 路径；3.4 节介绍 OSPFv2 特殊选项的配置，包括 Hello 间隔和 Dead 间隔的修改、干预 DR 和 BDR 的选举以及干预 OSPFv2 选路等内容。本章设计了 7 个应用场景、5 个实验和 2 个挑战练习，帮助读者熟练掌握本章内容。

3.1 学习 OSPFv2

3.1.1 学习 OSPFv2 基础知识

1. OSPFv2 简介

① OSPF（Open Shortest Path First，开放式最短路径优先）是一种链路状态路由协议，属于内部网关协议（Interior Gateway Protocol，IGP），是为替代使用距离矢量算法的 RIP（Routing Information Protocol，路由信息协议）而开发的。

② OSPFv2 是 OSPF 第 2 个版本，与 RIP 相比，OSPFv2 具有巨大优势，因为它收敛速度快，适用于规模较大的网络。

③ OSPFv2 使用了区域概念，网络管理员可以把路由域划分为不同的区域，方便对路由更新流量实施控制。

④ OSPFv2 有 Hello、DBD、LSR、LSU 和 LSAck 五种基本数据包类型，其中 LSU（Link-State Update，链路状态更新）数据包最重要，一个 LSU 数据包中可包含一种或多种 LSA（Link-State Advertisement，链路状态通告）数据包。

⑤ OSPFv2 采用组播方式通信，组播地址为 224.0.0.5/224.0.0.6；OSPFv2 报头标识了区域 ID、路由器 ID 和封装的数据类型。

2. OSPFv2 数据库

OSPFv2 消息可用于创建和维护 3 个 OSPFv2 数据库，如表 3-1 所示。

① 邻接数据库：即邻居表。

② 链路状态数据库（LSDB）：即拓扑表。

③ 转发数据库：即路由表。

表 3-1　OSPFv2 数据库

数据库	说明
邻接数据库	● 列出了已经与本地路由器建立了双向通信关系的所有邻居路由器 ● 该表对于每个路由器都是唯一的 ● 可以使用 show ip ospf neighbor 命令查看邻居表
链路状态数据库	● 列出网络中所有其他路由器的相关信息 ● 该数据库用于构建网络拓扑 ● 某个区域内的所有路由器都有相同的 LSDB ● 可以使用 show ip ospf database 命令查看 LSDB
转发数据库	● 列出了 SPF 算法根据链路状态计算生成的路由数据库 ● 每台路由器上的路由表都是唯一的，其中包含了如何把数据包发送给其他路由器的信息 ● 可以使用命令 show ip route 来查看路由表

3．SPF 算法

SPF（最短路径优先）算法是 OSPF 路由协议的基础，该算法有时也被称为 Dijkstra 算法，这是因为该算法是 Dijkstra 发明的。路由器在构建拓扑表时，会使用 Dijkstra SPF 算法计算出结果。SPF 算法基于到达目的地的累计开销，将每台路由器置于树的根部并计算到达每个节点的最短路径，从而创建 SPF 树，然后使用 SPF 树计算最佳路由。OSPFv2 将最佳路由放入转发数据库，用于创建路由表。

3.1.2　学习 OSPFv2 的基本特征

路由协议实际上就是一些通信规则，而 OSPFv2 可以被认为是一种"视情况而定"的规则。当网络处于相对稳定状态时，OSPFv2 将消耗较少的流量。通常，OSPFv2 每 30 min 向网内邻居路由器发送路由更新信息；当网络突然发生变化时，OSPFv2 将泛洪更多的流量，因此，它必须立即向受到影响的每台路由器发送 LSA（Link State Advertisement，链路状态通告），从而使网络能再次快速收敛。

OSPFv2 是一种链路状态路由协议，链路状态协议需要消耗更多的 CPU 和系统内存资源。其主要原因之一是，由于链路状态协议是基于分布式的地图概念，也就是说每一台路由器都拥有一份定期更新的网络地图的副本。除此以外，链路状态协议的路由表大小、区域中路由器的数量以及路由器之间邻接关系的数量都将影响路由器内存和 CPU 的利用率。

OSPFv2 的特征如表 3-2 所示。

表 3-2　OSPFv2 的特征

特征编号	特征项	特征描述
1	收敛特性	收敛速度快，网络中不存在路由环路
2	可扩展性	支持分层网络设计，适用于大规模网络，可扩展性强
3	基于算法	采用 SPF 算法
4	有类/无类协议	支持 VLSM 和 CIDR，支持不连续网络，支持手动路由汇总，属于无类路由协议

续表

特征编号	特征项	特征描述
5	更新方式	当网络发生变化时，采用触发更新方式；当网络相对稳定时，采用每隔 30 min 定期更新一次完整的拓扑表的方式，确保链路状态数据库（LSDB）信息的同步，网络中无路由环路
6	负载均衡	支持等价负载均衡，默认最大支持 4 条路径负载均衡
7	身份验证	支持明文认证，支持消息摘要 MD5 认证
8	通信方式	采用组播通信方式，组播地址为 224.0.0.5 或 224.0.0.6
9	管理距离	默认管理距离（AD）为 110
10	度量标准	度量为开销，Cost=10^8/带宽（bps）
11	资源利用	由于频繁调用 SPF 算法，且该算法本身计算复杂度较高，因此对 CPU 和内存资源消耗较大
12	管理维护	对网络管理员要求较高，实施和维护比较复杂
13	维护数据库	邻居表、LSDB（链路状态数据库）、路由表（转发数据库）

3.1.3 学习 OSPFv2 的原理

为了维护路由信息，OSPFv2 路由器需要完成一个通用的链路状态路由过程，并达到收敛状态。路由器之间的每条链路上都有一个开销值，在 OSPFv2 中，开销是用来确定去往目的地的最优路径的。路由器会按照以下步骤完成链路状态路由过程。

1．建立邻居邻接关系

启用 OSPFv2 的路由器将 Hello 数据包从所有启用 OSPFv2 的接口发送出去，以确定这些链路上是否存在邻居。如果存在邻居，启用 OSPFv2 的路由器将尝试与该邻居建立邻接关系。

2．交换 LSA

建立邻接关系之后，路由器之间会交换链路状态通告（LSA）。LSA 包含每个直连链路的状态和开销，路由器将其 LSA 泛洪到其邻居。收到 LSA 的邻接邻居会立即将 LSA 泛洪到其他直连的邻居，直到区域中的所有路由器收到所有 LSA。

3．构建 LSDB

在收到 LSA 后，启用 OSPFv2 的路由器会根据收到的 LSA 来构建链路状态数据库（LSDB），该数据库中最终会包含与该区域有关的所有拓扑信息。

4．执行 SPF 算法

接着路由器会执行 SPF 算法，根据 SPF 算法结果创建 SPF 树。

5．选择最优路由

在 SPF 树创建完成后，路由器会把去往每个网络的最优路径放入 IP 路由表。除非有管理距离较短的静态路由，否则会将该路由插入路由表。系统将根据路由表中的条目做出路由决策。

3.1.4　学习 OSPFv2 的数据包类型

1．Hello 数据包

Hello 数据包主要用于与其他 OSPFv2 路由器建立和维持邻居关系，以及 DR/BDR 选举。OSPFv2 Hello 数据包的格式如图 3-1 所示，其主要字段的含义如下所述。

① 网络掩码：发送 OSPFv2 数据包接口所在网络的子网掩码。如果相邻两台路由器的网络掩码不同，则不能建立邻居关系。

图 3-1　OSPFv2 Hello 数据包的格式

② Hello 间隔：连续 2 次发送 Hello 数据包之间的时间间隔，单位为秒。如果相邻两台路由器的 Hello 间隔时间不同，则不能建立邻居关系。

③ 路由器优先级：用于 DR/BDR 选举，范围为 0～255。如果设置为 0，则该路由器接口不能成为 DR/BDR。

④ 路由器 Dead 间隔：宣告邻居路由器无效之前等待的最长时间，又称失效时间。如果在此时间内未收到邻居发来的 Hello 数据包，则认为邻居失效；如果相邻两台路由器的失效时间不同，则不能建立邻居关系。

⑤ 指定路由器（DR）：用 DR 的接口 IP 地址表示。

⑥ 备份指定路由器（BDR）：用 BDR 的接口 IP 地址表示。

⑦ 邻居列表：列出建立邻居关系的邻居路由器的 Router ID。

2. DBD（DataBase Description，数据库描述）数据包

两台路由器在进行数据库同步时，用 DBD 数据包来描述自己的 LSDB，其内容包括 LSDB 中每一条 LSA Header（LSA 头部），LSA Header 可以唯一标识一条 LSA。LSA Header 只占一条 LSA 的整个数据量的一小部分，这样可以减少路由器之间的数据流量，对端路由器根据 LSA Header 就可以判断出是否已有这条 LSA。OSPFv2 DBD 数据包格式如图 3-2 所示，主要字段含义如下所述。

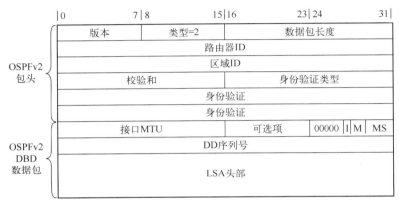

图 3-2　OSPFv2 DBD 数据包格式

① 接口 MTU：在数据包不分段的情况下，路由器接口能发送的最大 IP 数据包的大小。
② I：初始位，发送的第一个 DBD 数据包 I 位置 1，后续发送的 DBD 数据包 I 位置 0。
③ M：后继位，发送的最后一个 DBD 数据包的 M 位置 0，其他 DBD 数据包的 M 位置 1。
④ MS：主从位，用于协商主/从路由器，置 1 表示主路由器，置 0 表示从路由器。
⑤ DD 序列号：在数据库同步过程中，用来确保 DBD 数据包传输的可靠性和完整性，该序列号由主路由器在发送第一个 DBD 数据包时设置，后续数据包的序列号将依次增加。

3. LSR（Link-State Request，链路状态请求）数据包

在 LSDB 同步过程中，路由器在互相交换过 DBD 数据包之后，会查看本地 LSDB 中不包括哪些 LSA，或者对端 LSDB 中的哪些 LSA 比自己的更新，然后将这些更新的 LSA 记录在链路状态请求列表中，并通过发送 LSR 数据包来请求所需要的 LSA 条目的详细信息，内容包括所需要的 LSA 的摘要。OSPFv2 LSR 数据包格式如图 3-3 所示，主要字段的含义如下所述。

① 链路状态类型：LSA 的类型。
② 链路状态 ID：标识 LSA，LSA 类型不同标识方法也不同。
③ 通告路由器：始发 LSA 通告的路由器，表现为该路由器的路由器 ID。

图 3-3　OSPFv2 LSR 数据包格式

4. LSU（Link-State Update，链路状态更新）数据包

LSU 数据包用于回复 LSR 或通告新的 OSPFv2 更新，其内容是多条 LSA（全部内容）的集合。OSPFv2 LSU 数据包格式如图 3-4 所示，主要字段的含义如下所述。

① LSA 的数目：更新数据包中包含的 LSA 的数量。

② LSAs：更新数据包中包含的所有 LSA。

图 3-4　OSPFv2 LSU 数据包格式

5. LSAck（Link-State Acknowledgement，链路状态确认）数据包

LSAck 数据包用于对接收到的 LSU 数据包进行确认，其内容是需要确认的 LSA 头部。一个 LSAck 数据包可对多个 LSA 进行确认。OSPFv2 LSAck 数据包格式如图 3-5 所示。

说明：

① LSA 头部：该数据包包含的 LSA 头部。

② OSPFv2 的 LSU 中可以包含一个或多个不同类型的 LSA。

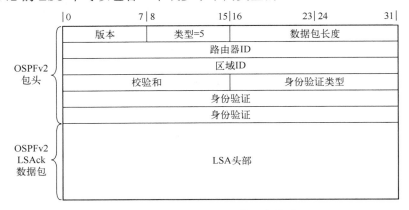

图 3-5　OSPFv2 LSAck 数据包格式

3.1.5　学习 OSPFv2 的运行状态

当将 OSPFv2 路由器初次连接到网络时，它会尝试以下操作：

- 与邻居建立邻接关系；
- 交换路由信息；
- 计算最佳路由；
- 实现收敛。

为了达到收敛状态，OSPFv2 会经历以下几种状态。

1．Down 状态

描述：Down 状态是一种失效状态，在 Down 状态下，OSPFv2 进程还没有与任何邻居交换信息，具体描述如下。

- 没有收到 Hello 数据包；
- 路由器发送 Hello 数据包；
- 将过渡到 Init 状态。

2．Init 状态

描述：在 Init 状态下，OSPFv2 路由器以固定的时间间隔发送 Hello 数据包，具体描述如下。

- OSPFv2 路由器发送 Hello 数据包；
- Hello 数据包中包含有发送方路由器的路由器 ID；
- 将过渡到 Two-Way 状态。

3. Two-Way 状态

描述：在 Two-Way 状态下，OSPFv2 路由器间建立双向通信，具体描述如下。
- 在这种状态下，这两台路由器之间的通信是双向的；
- 在多路访问链路上，OSPFv2 路由器之间会选举出 DR 和 BDR；
- 将过渡到 ExStart 状态。

4. ExStart 状态

描述：ExStart 状态是信息交换初始化状态，具体描述如下。
- OSPFv2 邻居双方会发起数据库同步过程，选举出主/从路由器；
- 在点对点网络中，两台 OSPFv2 路由器会确定由哪台路由器来初始化 DBD 数据包交换的过程，并且会确定初始 DBD 数据包的序列号。

5. Exchange 状态

描述：Exchange 状态是信息交换状态，具体描述如下。
- OSPFv2 路由器间交换 DBD 数据包；
- 如果 OSPFv2 路由器还要请求更多的路由器信息，则进入 Loading 状态，否则进入为 Full 状态。

6. Loading 状态

描述：Loading 状态是信息加载状态，OSPFv2 邻居间同步数据库，具体描述如下。
- OSPFv2 路由器通过 LSR 和 LSU 获取更多路由信息；
- OSPFv2 路由器使用 SPF 算法进行处理；
- 将过渡到 Full 状态。

7. Full 状态

描述：Full 状态是完全邻接状态，OSPFv2 邻居已建立完全邻接系统。
- OSPFv2 路由器的链路状态数据库已经完全同步。

3.1.6 学习 OSPFv2 路由器类型

OSPFv2 使用 4 种不同类型的路由器来构建分层区域结构。OSPFv2 路由器的类型决定了什么样的数据流能够进入和离开区域。在这种区域分层结构中，每台路由器都担任相应的角色。一个典型的 OSPFv2 网络如图 3-6 所示，在该网络中涵盖了所有 4 种类型的 OSPFv2 路由器，下面我们将对这 4 种类型的路由器进行简要介绍。

1. 内部路由器（IR）

内部路由器（Interior Router，IR）是所有接口都位于同一个区域中的路由器。内部路由器（IR）有骨干内部路由器和非骨干内部路由器两种。该类型的路由器只有一个链路状态数据库

（LSDB）数据。

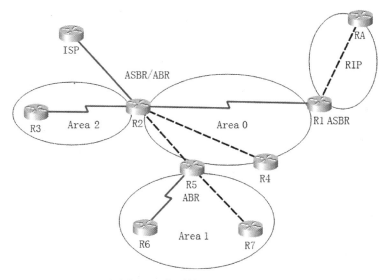

图 3-6　典型的 OSPFv2 网络

2. 骨干路由器（BR）

骨干路由器（Backbone Router，BR）是至少有一个接口与骨干区域 0（Area 0）相连的路由器。该类型路由器可能所有接口都落在 Area 0 内，属于内部路由器，即骨干内部路由器；该类型的路由器也可能与非骨干区域相连。

3. 区域边界路由器（ABR）

区域边界路由器（Area Border Router，ABR）是接口至少属于两个区域的路由器，并且两个区域中有一个区域是骨干区域。该类型路由器与多个区域相连，对于连接的每个区域，路由器都有一个独立的链路状态数据库。ABR 会将每个区域的链路状态数据库汇总成 Summary LSA，然后转发至骨干区域以便分发到其他区域。所以，通常 ABR 既属于骨干区域，又属于非骨干区域。

4. 自治系统边界路由器（ASBR）

自治系统（Autonomous System，AS）：采用同一种路由协议交换路由信息的路由器及其网络构成一个自治系统。

自治系统边界路由器（Autonomous System Border Router，ASBR）是至少有一个接口与其他 AS（运行非 OSPFv2 的路由区域）的路由器相连且互相交换路由信息的路由器。ASBR 可以连接多个 AS，ASBR 负责将通过交互学到的外部路由信息通告给它所连接的 AS。ASBR 可能同时运行 OSPFv2 和其他路由协议，如 RIP 或 BGP 等，ASBR 负责处理外部路由。

在 OSPFv2 网络中，同一台路由器可属于多种类型，比如可能既是 ABR，同时又是 ASBR。

为了使读者熟练辨别 OSPFv2 网络中的 4 种路由器类型，现将图 3-6 中的路由器类型进行归类，具体如表 3-3 所示。

表 3-3 OSPFv2 网络中的 4 种路由器类型

路由器类型编号	路由器类型	路由器名称
1	骨干路由器（BR）	R1、R2、R4、R5
2	内部路由器（IR）	R3、R4、R6、R7
3	区域边界路由器（ABR）	R2、R5
4	自治系统边界路由器（ASBR）	R1、R2

3.2 配置基础 OSPFv2

OSPFv2 的配置语法如下：

> Router(config)#**router ospf** *process-id*
> Router(config-router)#**network** *network-address wildcard-mask* **area** *area-id*
> //在指定接口上启用 OSPFv2 进程并宣告

配置 OSPFv2 步骤如下：

① 启用 OSPFv2 进程。进程号取值范围为 1～65535，进程号只在本地有效，每台路由器可支持多个 OSPFv2 进程。执行命令 **no router ospf** *process-id* 关闭 OSPFv2 进程。

② 用 **network** 命令指定参与 OSPFv2 进程的接口。*network-address* 可以是主网地址、子网地址，也可以是接口地址。通过通配符掩码匹配指定接口。

3.2.1 认识 OSPFv2 路由器 ID

OSPFv2 路由器 ID 长 32 位，与 IPv4 地址的格式相同。OSPFv2 路由器 ID 的作用是唯一地标识一台 OSPFv2 路由器。OSPFv2 所有类型的数据包中都包含始发路由器的 OSPFv2 路由器 ID。OSPFv2 路由器 ID 可以由管理员定义，也可以由路由器自动分配。启用了 OSPFv2 的路由器会在以下情景中使用 OSPFv2 路由器 ID。

- 参与 OSPFv2 数据库的同步：在 Exchange 状态中，OSPFv2 路由器 ID 最高的路由器会首先发送自己的数据库描述数据包（DBD 数据包）。
- 参与指定路由器（DR）的选举：在多路访问 LAN 环境中，默认情况下，OSPFv2 路由器 ID 最高的路由器会被选为 DR。OSPFv2 路由器 ID 次高的路由器会被选举为 BDR。
注意：后文会更详细地讨论 DR 和 BDR 的选举过程。

3.2.2 场景一：配置 OSPFv2 路由器 ID

经过 3.2.1 节的学习，相信大家已经对路由器 ID 有了初步认知。接下来，让我们一起通过具体实验来配置路由器 ID。配置路由器 ID 实验拓扑如图 3-7 所示。

图 3-7 配置路由器 ID 实验拓扑

步骤一：配置路由器 R1 并启用 OSPFv2 进程

```
R1(config)#interface GigabitEthernet 0/0/0
R1(config-if)#ip address 192.168.12.1 255.255.255.0
R1(config-if)#no shutdown
R1(config-if)#interface Loopback 0
R1(config-if)#ip address 192.168.1.1 255.255.255.255
R1(config-if)#router ospf 1
R1(config-router)#router-id 1.1.1.1
```

步骤二：配置路由器 R2 并启用 OSPFv2 进程

```
R2(config)#interface GigabitEthernet 0/0/0
R2(config-if)#ip address 192.168.12.2 255.255.255.0
R2(config-if)#no shutdown
R2(config-if)#interface Loopback 0
R2(config-if)#ip address 192.168.2.1 255.255.255.255
R2(config-if)#router ospf 1
```

步骤三：查看路由器 R1 和 R2 的路由器 ID

```
R1#show ip protocols
Routing Protocol is "ospf 1"
  Outgoing update filter list for all interfaces is not set
  Incoming update filter list for all interfaces is not set
  Router ID 1.1.1.1
  Number of areas in this router is 0. 0 normal 0 stub 0 nssa
  Number of areas in this router is 0. 0 normal 0 stub 0 nssa
```

以上输出结果表明，路由器 R1 的路由器 ID 为采用 **router-id** 命令手动指定的 IP 地址。

```
R2#show ip protocols
Routing Protocol is "ospf 1"
  Outgoing update filter list for all interfaces is not set
  Incoming update filter list for all interfaces is not set
  Router ID 192.168.2.1
```

以上输出结果表明，路由器 R2 的路由器 ID 为环回接口的 IP 地址。假设没有环回接口，路由器 ID 则会选择活动物理接口的最大 IP 地址。

由此可知，路由器 ID 优先选择采用命令手动指定的 IP 地址，其次选择环回接口的最大 IP 地址，最后才会选择活动物理接口的最大 IP 地址。

3.2.3 认识单区域 OSPFv2

当网络中的路由器数量较少时，我们可以不对其进行多区域划分，让其只运行在一个区域中，即 Area 0，就是单区域。在单区域中：
- 路由器的所有接口都属于一个区域，适合小型网络。
- 所有路由器可以划分在骨干区域，也可以全部划分在非骨干区域；前者可以扩展为多区域，后者不便于扩展为多区域，最佳区域号为 0。
- 所有学习来的路由都是域内路由。
- 单区域的所有路由器都是内部路由器，如果在骨干区域，则承担内部路由器兼骨干路由器双重角色；如果在非骨干区域，则承担内部路由器的角色。

3.2.4 场景二：配置单区域 OSPFv2

经过 3.2.3 节的学习，相信大家已经初步认识了单区域 OSPFv2。接下来，让我们一起通过具体实验来配置单区域 OSPFv2。配置单区域 OSPFv2 实验拓扑如图 3-8 所示。

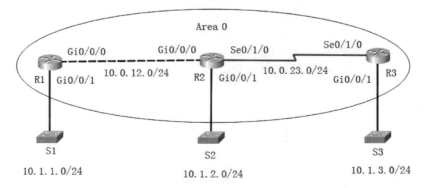

图 3-8　配置单区域 OSPFv2 实验拓扑

步骤一：为设备命名并配置接口 IP 地址

（1）在路由器 R1 上为设备命名并配置接口 IP 地址

```
Router(config)#hostname R1
R1(config)#interface GigabitEthernet0/0/0
R1(config-if)#ip address 10.0.12.1 255.255.255.0
R1(config-if)#no shutdown
R1(config-if)#interface GigabitEthernet0/0/1
R1(config-if)#ip address 10.1.1.254 255.255.255.0
R1(config-if)#no shuttdown
```

（2）在路由器 R2 上为设备命名并配置接口 IP 地址

```
Router(config)#hostname R2
R2(config)#interface GigabitEthernet0/0/0
R2(config-if)#ipaddress 10.0.12.2 255.255.255.0
R2(config-if)#no shutdown
R2(config-if)#interface GigabitEthernet0/0/1
R2(config-if)#ip address 10.1.2.254 255.255.255.0
R2(config-if)# no shutdown
R2(config)#interface Serial0/1/0
R2(config-if)#ip address 10.0.23.2 255.255.255.0
R2(config-if)# no shutdown
```

（3）在路由器 R3 上为设备命名并配置接口 IP 地址

```
Router(config)#hostname R3
R3(config)#interface Serial 0/1/0
R3(config-if)#ip address 10.0.23.3 255.255.255.0
R2(config-if)# no shutdown
R3(config)#interface GigabitEthernet0/0/1
R3(config-if)#ip address 10.1.3.254 255.255.255.0
R2(config-if)# no shutdown
```

步骤二：配置单区域 OSPFv2

（1）配置路由器 R1

```
R1(config)#router ospf 1
R1(config-router)#router-id 1.1.1.1
```

R1(config-router)#**network 10.1.1.0 0.0.0.255 area 0**

R1(config-router)#**network 10.0.12.0 0.0.0.255 area 0**   //R1 为内部路由器（IR）和骨干路由器（BR）

（2）配置路由器 R2

R2(config)#**router ospf 1**

R2(config-router)#**router-id 2.2.2.2**

R2(config-router)#**network 10.1.2.0 0.0.0.255 area 0**

R2(config-router)#**network 10.0.12.0 0.0.0.255 area 0**

R2(config-router)#**network 10.0.23.0 0.0.0.255 area 0**   //R2 为内部路由器（IR）和骨干路由器（BR）

（3）配置路由器 R3

R3(config)#**router ospf 1**

R3(config-router)#**router-id 3.3.3.3**

R3(config-router)#**network 10.0.23.0 0.0.0.255 area 0**

R3(config-router)#**network 10.1.3.0 0.0.0.255 area 0**   //R3 为内部路由器（IR）和骨干路由器（BR）

步骤三：查看路由器的路由表

（1）查看路由器 R1 的路由表

```
R1# show ip route | begin 10.0.0.0
     10.0.0.0/8 is variably subnetted, 7 subnets, 2 masks
C       10.0.12.0/24 is directly connected, GigabitEthernet0/0/0
L       10.0.12.1/32 is directly connected, GigabitEthernet0/0/0
O       10.0.23.0/24 [110/65] via 10.0.12.2, 00:04:55, GigabitEthernet0/0/0      //域内路由
C       10.1.1.0/24 is directly connected, GigabitEthernet0/0/1
L       10.1.1.254/32 is directly connected, GigabitEthernet0/0/1
O       10.1.2.0/24 [110/2] via 10.0.12.2, 00:05:05, GigabitEthernet0/0/0        //域内路由
O       10.1.3.0/24 [110/66] via 10.0.12.2, 00:03:06, GigabitEthernet0/0/0       //域内路由
```

（2）查看路由器 R2 的路由表

```
R2# show ip route | begin 10.0.0.0
     10.0.0.0/8 is variably subnetted, 8 subnets, 2 masks
C       10.0.12.0/24 is directly connected, GigabitEthernet0/0/0
L       10.0.12.2/32 is directly connected, GigabitEthernet0/0/0
C       10.0.23.0/24 is directly connected, Serial0/1/0
L       10.0.23.2/32 is directly connected, Serial0/1/0
O       10.1.1.0/24 [110/2] via 10.0.12.1, 00:05:44, GigabitEthernet0/0/0        //域内路由
```

	C	10.1.2.0/24 is directly connected, GigabitEthernet0/0/1	
	L	10.1.2.254/32 is directly connected, GigabitEthernet0/0/1	
	O	**10.1.3.0/24 [110/65] via 10.0.23.3, 00:03:48, Serial0/1/0**	//域内路由

（3）查看路由器 R3 的路由表

R3# **show ip route | begin 10.0.0.0**
    10.0.0.0/8 is variably subnetted, 7 subnets, 2 masks

	O	**10.0.12.0/24 [110/65] via 10.0.23.2, 00:04:25, Serial0/1/0**	//域内路由
	C	10.0.23.0/24 is directly connected, Serial0/1/0	
	L	10.0.23.3/32 is directly connected, Serial0/1/0	
	O	**10.1.1.0/24 [110/66] via 10.0.23.2, 00:04:25, Serial0/1/0**	//域内路由
	O	**10.1.2.0/24 [110/65] via 10.0.23.2, 00:04:25, Serial0/1/0**	//域内路由
	C	10.1.3.0/24 is directly connected, GigabitEthernet0/0/1	
	L	10.1.3.254/32 is directly connected, GigabitEthernet0/0/1	

步骤四：测试网络连通性

（1）在路由器 R1 上访问 10.1.2.0/24 网段

R1#**ping 10.1.2.254**
Sending 5, 100-byte ICMP Echos to 10.1.2.254, timeout is 2 seconds:
!!!!!
Success rate is 100 percent (5/5), round-trip min/avg/max = 0/0/2 ms

（2）在路由器 R1 上访问 10.1.3.0/24 网段

R1#**ping 10.1.3.254**
Sending 5, 100-byte ICMP Echos to 10.1.3.254, timeout is 2 seconds:
!!!!!
Success rate is 100 percent (5/5), round-trip min/avg/max = 1/6/13 ms

至此，单区域 OSPFv2 配置实验已经完成。单区域 OSPFv2 所有路由器的角色都是内部路由器，所学习到的 OSPF 路由都是域内路由。

3.2.5 认识多区域 OSPFv2

一个区域中有过多的路由器会使 LSDB 非常庞大并增加 CPU 的负载。因此，将路由器有效分区，将巨大的数据库分成更小、更易管理的数据库是很有必要的。

多区域 OSPFv2 分层拓扑设计的优势如下所述。

① 缩小了路由表：由于在区域之间可以对网络地址进行汇总，路由表条目比较少，所以，路由表比较小。默认情况下不启用路由汇总。

② 减少了链路状态更新的负载：通过将较大的 OSPFv2 区域划分为多个较小的 OSPFv2 区域，可以降低对处理器和内存需求。

③ 降低了 SPF 计算的频率：在多区域 OSPFv2 中，拓扑的变化只会影响一个区域内部，因此降低了 SPF 计算的频率。例如，由于 LSA 泛洪在区域边界终止，这使得路由更新的影响降到最小。

注意：
- 学习来的路由包括域内路由（本区域路由）和域间路由（其他区域路由）；
- 常规区域必须与骨干区域直接相连；
- 网络中至少存在一个 ABR，且一定存在骨干路由器。

多区域 OSPFv2 网络拓扑如图 3-9 所示，在该图中，R2 是 Area 2 中的 ABR，Area 2 中网络拓扑的变化会导致 Area 2 中的所有路由器都要重新执行 SPF 计算，创建出新的 SPF 树并更新其 IP 路由表。接着 ABR R2 会向 Area 0 中的路由器发送 LSA，最终该 LSA 会被泛洪到 OSPFv2 路由域中的所有路由器。但这种类型的 LSA 不会导致其他区域中的路由器重新执行 SPF 计算，路由器只需更新本地 LSDB 和路由表。从图 3-9 中可知，多区域 OSPFv2 把故障隔离在一个区域内，节省了资源。Area 2 中的故障让区域内的路由器之间交换 LSA 并执行 SPF 计算，但这不会影响 Area 0 和 Area 1，所以说链路更改仅影响本地区域。

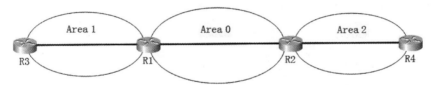

图 3-9　多区域 OSPFv2 网络拓扑

3.2.6　场景三：配置多区域 OSPFv2

经过 3.2.5 节的学习，相信大家已经了解到了多区域 OSPFv2 的相关知识。接下来，让我们一起通过具体实验来配置多区域的 OSPFv2，配置多区域的 OSPFv2 实验拓扑如图 3-10 所示。

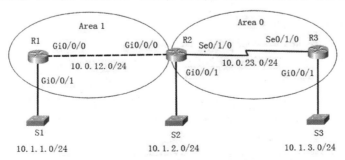

图 3-10　配置多区域的 OSPFv2 实验拓扑

步骤一：为设备命名并配置接口地址

路由器 R1、R2 和 R3 的基础配置与 3.2.4 节中场景二的步骤一相同。

步骤二：配置多区域 OSPFv2

（1）配置路由器 R1

```
R1(config)#router ospf 1
R1(config-router)#router-id 1.1.1.1
R1(config-router)#network 10.1.1.0 0.0.0.255 area 1
R1(config-router)#network 10.0.12.0 0.0.0.255 area 1      //R1 为内部路由器（IR）
```

（2）配置路由器 R2

```
R2(config)#router ospf 1
R2(config-router)#router-id 2.2.2.2
R2(config-router)#network 10.0.12.0 0.0.0.255 area 1
R2(config-router)#network 10.1.2.0 0.0.0.255 area 0
R2(config-router)#network 10.0.23.0 0.0.0.255 area 0      //R2 为区域边界路由器（ABR）
```

（3）配置路由器 R3

```
R3(config)#router ospf 1
R3(config-router)#router-id 3.3.3.3
R3(config-router)#network 10.1.3.0 0.0.0.255 area 0
R3(config-router)#network 10.0.23.0 0.0.0.255 area 0      //R3 为内部路由器（IR）和骨干路由器（BR）
```

步骤三：查看路由表

（1）查看路由器 R1 的路由表

```
R1# show ip route | begin 10.0.0.0
     10.0.0.0/8 is variably subnetted, 7 subnets, 2 masks
C       10.0.12.0/24 is directly connected, GigabitEthernet0/0/0
L       10.0.12.1/32 is directly connected, GigabitEthernet0/0/0
O IA    10.0.23.0/24 [110/65] via 10.0.12.2, 02:41:48, GigabitEthernet0/0/0    //域间路由
C       10.1.1.0/24 is directly connected, GigabitEthernet0/0/1
L       10.1.1.254/32 is directly connected, GigabitEthernet0/0/1
O IA    10.1.2.0/24 [110/2] via 10.0.12.2, 02:41:48, GigabitEthernet0/0/0      //域间路由
O IA    10.1.3.0/24 [110/66] via 10.0.12.2, 02:38:50, GigabitEthernet0/0/0     //域间路由
```

（2）查看路由器 R2 的路由表

```
R2# show ip route | begin 10.0.0.0
        10.0.0.0/8 is variably subnetted, 8 subnets, 2 masks
C       10.0.12.0/24 is directly connected, GigabitEthernet0/0/0
L       10.0.12.2/32 is directly connected, GigabitEthernet0/0/0
C       10.0.23.0/24 is directly connected, Serial0/1/0
L       10.0.23.2/32 is directly connected, Serial0/1/0
O       10.1.1.0/24 [110/2] via 10.0.12.1, 01:42:01, GigabitEthernet0/0/0    //域内路由
C       10.1.2.0/24 is directly connected, GigabitEthernet0/0/1
L       10.1.2.254/32 is directly connected, GigabitEthernet0/0/1
O       10.1.3.0/24 [110/65] via 10.0.23.3, 01:41:56, Serial0/1/0            //域内路由
```

（3）查看路由器 R3 的路由表

```
R3# show ip route | begin 10.0.0.0
        10.0.0.0/8 is variably subnetted, 7 subnets, 2 masks
O IA    10.0.12.0/24 [110/65] via 10.0.23.2, 01:42:36, Serial0/1/0           //域间路由
C       10.0.23.0/24 is directly connected, Serial0/1/0
L       10.0.23.3/32 is directly connected, Serial0/1/0
O IA    10.1.1.0/24 [110/66] via 10.0.23.2, 01:42:36, Serial0/1/0            //域间路由
O       10.1.2.0/24 [110/65] via 10.0.23.2, 01:42:36, Serial0/1/0            //域内路由
C       10.1.3.0/24 is directly connected, GigabitEthernet0/0/1
L       10.1.3.254/32 is directly connected, GigabitEthernet0/0/1
```

步骤四：测试网络连通性

（1）在路由器 R1 上访问 10.1.2.0/24 网段

```
R1#ping 10.1.2.254
Sending 5, 100-byte ICMP Echos to 10.1.2.254, timeout is 2 seconds:
!!!!!
Success rate is 100 percent (5/5), round-trip min/avg/max = 0/0/1 ms
```

（2）在路由器 R1 上访问 10.1.3.0/24 网段

```
R1#ping 10.1.3.254
Sending 5, 100-byte ICMP Echos to 10.1.3.254, timeout is 2 seconds:
!!!!!
Success rate is 100 percent (5/5), round-trip min/avg/max = 5/8/11 ms
```

至此，多区域 OSPFv2 配置实验已经完成。多区域 OSPFv2 路由器的角色有骨干路由器、内部路由器和区域边界路由器，OSPF 路由类型有域内路由和域间路由。

3.3 检验 OSPFv2 配置

3.3.1 实验一：检验 OSPFv2 配置

检验 OSPFv2 配置实验拓扑如图 3-11 所示，以路由器 R2 为例。

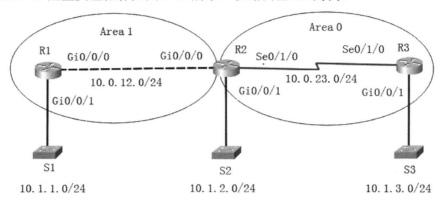

图 3-11　检验 OSPFv2 配置实验拓扑

（1）采用 **show running-config** 命令检验 OSPFv2 配置

> **R2#show running-config | section router ospf 1**
> **router ospf 1**
> 　router-id 2.2.2.2
> 　log-adjacency-changes
> 　**network 10.1.2.0 0.0.0.255 area 0**
> 　**network 10.0.23.0 0.0.0.255 area 0**
> 　**network 10.0.12.0 0.0.0.255 area 1**

（2）采用 **show ip protocols** 命令检验 OSPFv2 配置

> **R2>show ip protocols**
> **Routing Protocol is "ospf 1"**
> 　Outgoing update filter list for all interfaces is not set
> 　Incoming update filter list for all interfaces is not set
> 　**Router ID 2.2.2.2**
> 　Number of areas in this router is 2. 2 normal 0 stub 0 nssa
> 　Maximum path: 4
> 　Routing for Networks:
> 　　**10.1.2.0 0.0.0.255 area 0**

        10.0.23.0 0.0.0.255 area 0
        10.0.12.0 0.0.0.255 area 1
    Routing Information Sources:
        Gateway   Distance   Last Update
        1.1.1.1   110   00:01:11
        2.2.2.2   110   00:01:11
        3.3.3.3   110   00:01:45
    Distance: (default is 110)

（3）采用 **show ip ospf** [*process-id*]命令查看 OSPFv2 进程信息

```
R2>show ip ospf 1
    Routing Process "ospf 1" with ID 2.2.2.2
    Supports only single TOS(TOS0) routes
    Supports opaque LSA
    It is an area border router          //R2 是一台区域边界路由器
    SPF schedule delay 5 secs, Hold time between two SPFs 10 secs
    Minimum LSA interval 5 secs. Minimum LSA arrival 1 secs
    Number of external LSA 0. Checksum Sum 0x000000
    Number of opaque AS LSA 0. Checksum Sum 0x000000
    Number of DCbitless external and opaque AS LSA 0
    Number of DoNotAge external and opaque AS LSA 0
    Number of areas in this router is 2.2 normal 0 stub 0 nssa
    External flood list length 0
        Area BACKBONE(0)
            Number of interfaces in this area is 2       //区域 0 内接口数量为 2
            Area has no authentication                   //区域没有配置认证
            SPF algorithm executed 4 times               //SPF 算法执行了 4 次
            Area ranges are
            Number of LSA 4. Checksum Sum 0x01dc2d
            Number of opaque link LSA 0. Checksum Sum 0x000000
            Number of DCbitless LSA 0
            Number of indication LSA 0
            Number of DoNotAge LSA 0
            Flood list length 0
        Area 1
            Number of interfaces in this area is 1       //区域 1 内接口数量为 1
            Area has no authentication
            SPF algorithm executed 4 times               //SPF 算法执行了 4 次
            Area ranges are
            Number of LSA 6. Checksum Sum 0x02eb2e
```

```
                        Number of opaque link LSA 0. Checksum Sum 0x000000
                        Number of DCbitless LSA 0
                        Number of indication LSA 0
                        Number of DoNotAge LSA 0
                        Flood list length 0
```

（4）采用 **show ip ospf interface** [*interface*]命令查看 OSPFv2 接口信息

```
        R2>show ip ospf interface GigabitEthernet 0/0/0
        GigabitEthernet0/0/0 is up, line protocol is up
            Internet address is 10.0.12.2/24, Area 1
            Process ID 1, Router ID 2.2.2.2, Network Type BROADCAST, Cost: 1
            Transmit Delay is 1 sec, State DR, Priority 1
            Designated Router (ID) 2.2.2.2, Interface address 10.0.12.2
            Backup Designated Router (ID) 1.1.1.1, Interface address 10.0.12.1
            Timer intervals configured, Hello 10, Dead 40, Wait 40, Retransmit 5
                Hello due in 00:00:04
            Index 3/3, flood queue length 0
            Next 0x0(0)/0x0(0)
            Last flood scan length is 1, maximum is 1
            Last flood scan time is 0 msec, maximum is 0 msec
            Neighbor Count is 1, Adjacent neighbor count is 1
                Adjacent with neighbor 1.1.1.1    (Backup Designated Router)
            Suppress hello for 0 neighbor(s)
```

至此，检验 OSPFv2 配置实验结束。请读者熟练掌握命令 **show running-config**、**show ip protocols**、**show ip ospf** 以及 **show ip ospf interface**，并以此进行 OSPFv2 故障排错。

3.3.2 实验二：查看 OSPFv2 邻居表

查看 OSPFv2 邻居表实验拓扑如图 3-12 所示。

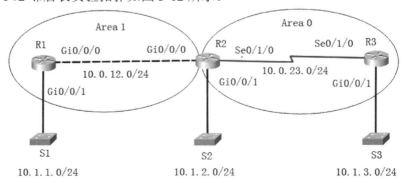

图 3-12 查看 OSPFv2 邻居表实验拓扑

（1）查看路由器 R1 的邻居表

R1#show ip ospf neighbor					
Neighbor ID	Pri	State	Dead Time	Address	Interface
2.2.2.2	1	FULL/DR	00:00:33	10.0.12.2	GigabitEthernet0/0/0

由以上输出可知，R1 通过优先级为 1 的 GigabitEthernet0/0/0 接口发现邻居 R2（路由器 ID 为 2.2.2.2），R2 接口的 IP 地址为 10.0.12.2，R2 的角色是 DR，R2 与 R1 的关系为 Full（完全邻接）。

（2）查看路由器 R2 的邻居表

R2#show ip ospf neighbor					
Neighbor ID	Pri	State	Dead Time	Address	Interface
1.1.1.1	1	FULL/BDR	00:00:38	10.0.12.1	GigabitEthernet0/0/0
3.3.3.3	0	FULL/ -	00:00:35	10.0.23.3	Serial0/1/0

由以上输出可知，R2 发现邻居 R1，R1 的角色是 BDR，R1 与 R2 关系为 Full。R2 通过优先级为 0 的 Serial0/1/0 接口发现邻居 R3，R3 与 R2 关系为 Full，它们之间不选举 DR。

（3）查看路由器 R3 的邻居表

R3#show ip ospf neighbor					
Neighbor ID	Pri	State	Dead Time	Address	Interface
2.2.2.2	0	FULL/ -	00:00:31	10.0.23.2	Serial0/1/0

至此，查看 OSPFv2 邻居表实验结束。请读者自行验证，相同网络拓扑，不论是 OSPF 单区域还是多区域，同一台路由器的邻居表是一样的。

3.3.3 实验三：查看 OSPFv2 拓扑表

查看单区域和多区域 OSPFv2 拓扑表实验拓扑分别如图 3-13（a）和图 3-13（b）所示。

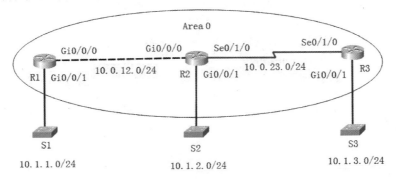

图 3-13（a） 查看单区域 OSPFv2 拓扑表实验拓扑

（1）查看单区域 OSPFv2 拓扑表

```
R1#show ip ospf database
        OSPF Router with ID (1.1.1.1) (Process ID 1)   //R1 是域内路由器（IR）和骨干路由器（BR）

            Router Link States (Area 0)

Link ID         ADV Router      Age         Seq#            Checksum Link count
1.1.1.1         1.1.1.1         1447        0x8000000b      0x00e016   2    //R1 有 2 条链路
3.3.3.3         3.3.3.3         1486        0x8000000b      0x0096a1   3    //R3 有 3 条链路
2.2.2.2         2.2.2.2         1449        0x8000000d      0x0011ea   4    //R2 有 4 条链路

            Net Link States (Area 0)
Link ID         ADV Router      Age         Seq#            Checksum
10.0.12.2       2.2.2.2         1449        0x80000008      0x00497b        //R2 为 DR
```

由以上输出可知，R1 收到 R1、R3 和 R2 有关区域 0 的路由器链路状态通告（路由器 LSA）。因为 R2 与 R3 之间有一条串行链路（由 DCE 线缆和 DTE 线缆连接而成），所以 R2 的链路数量为 4，R3 的链路数量为 3。R1 收到来自 DR R2（2.2.2.2）的网络状态通告（网络 LSA）。

```
R2#show ip ospf database
        OSPF Router with ID (2.2.2.2) (Process ID 1)

            Router Link States (Area 0)   //R2 是域内路由器（IR）和骨干路由器（BR）

Link ID         ADV Router      Age         Seq#            Checksum Link count
2.2.2.2         2.2.2.2         1493        0x8000000d      0x0011ea   4    //R2 有 4 条链路
3.3.3.3         3.3.3.3         1531        0x8000000b      0x0096a1   3    //R3 有 3 条链路
1.1.1.1         1.1.1.1         1492        0x8000000b      0x00e016   2    //R1 有 2 条链路

            Net Link States (Area 0)
Link ID         ADV Router      Age         Seq#            Checksum
10.0.12.2       2.2.2.2         1493        0x80000008      0x00497b        //R2 为 DR
```

由以上输出可知，R2 与同区域的路由器 R1 拥有相同的拓扑表。

```
R3#show ip ospf database
        OSPF Router with ID (3.3.3.3) (Process ID 1)
            Router Link States (Area 0)   //R3 是域内路由器（IR）和骨干路由器（BR）

Link ID         ADV Router      Age         Seq#            Checksum Link count
3.3.3.3         3.3.3.3         1577        0x8000000b      0x0096a1   3    //R3 有 3 条链路
2.2.2.2         2.2.2.2         1540        0x8000000d      0x0011ea   4    //R2 有 4 条链路
1.1.1.1         1.1.1.1         1539        0x8000000b      0x00e016   2    //R1 有 2 条链路

            Net Link States (Area 0)
Link ID         ADV Router      Age         Seq#            Checksum
10.0.12.2       2.2.2.2         1540        0x80000008      0x00497b        //R2 为 DR
```

由以上输出可知，路由器 R3 与 R1 和 R2 拥有相同的拓扑表。

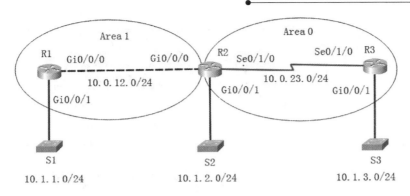

图 3-13（b） 查看多区域 OSPFv2 拓扑表实验拓扑

（2）查看多区域 OSPFv2 拓扑表

```
R1#show ip ospf database
        OSPF Router with ID (1.1.1.1) (Process ID 1)
            Router Link States (Area 1)       //R1 是域内路由器（IR）
Link ID     ADV Router      Age     Seq#        Checksum  Link count
1.1.1.1     1.1.1.1         364     0x8000000b  0x00e016  2    //R1 有 2 条链路在 Area 1 中
2.2.2.2     2.2.2.2         364     0x80000009  0x0034d7  1    //R2 有 1 条链路在 Area 1 中
            Net Link States (Area 1)
Link ID     ADV Router      Age     Seq#        Checksum
10.0.12.2   2.2.2.2         64      0x80000008  0x001f94
            Summary Net Link States (Area 1)
Link ID     ADV Router      Age     Seq#        Checksum
10.1.2.0    2.2.2.2         359     0x80000016  0x0088ab
10.0.23.0   2.2.2.2         359     0x80000017  0x0023bc
10.1.3.0    2.2.2.2         359     0x80000018  0x00fcf3
```

由以上输出可知，R1 收到 R1 和 R2 有关区域 1 的路由器 LSA，R1 收到来自 DR R2（2.2.2.2）的网络 LSA，R1 还收到来自 ABR R2（2.2.2.2）通告的区域 0 的网络汇总 LSA。

```
R2#show ip ospf database
        OSPF Router with ID (2.2.2.2) (Process ID 1)
            Router Link States (Area 0)       //R2 是骨干路由器（BR）
Link ID     ADV Router      Age     Seq#        Checksum  Link count
2.2.2.2     2.2.2.2         431     0x8000000c  0x001329  3    //R2 在 Area 0 中有 3
条链路
3.3.3.3     3.3.3.3         432     0x8000000b  0x0096a1  3    //R3 在 Area 0 中有 3
条链路
```

```
                    Summary Net Link States (Area 0)
        Link ID         ADV Router      Age         Seq#            Checksum
        10.0.12.0       2.2.2.2         391         0x8000000f      0x0034fd
        10.1.1.0        2.2.2.2         391         0x80000010      0x00a990
                    Router Link States (Area 1)       //R2 是区域边界路由器（ABR）
        Link ID         ADV Router      Age         Seq#            Checksum Link count
        2.2.2.2         2.2.2.2         395         0x80000009      0x0034d7  1    //R2 在 Area 1 中有 1
条链路
        1.1.1.1         1.1.1.1         396         0x8000000b      0x00e016  2    //R1 在 Area 1 中有 2
条链路
                    Net Link States (Area 1)
        Link ID         ADV Router      Age         Seq#            Checksum
        10.0.12.2       2.2.2.2         395         0x80000008      0x001f94
                    Summary Net Link States (Area 1)
        Link ID         ADV Router      Age         Seq#            Checksum
        10.1.2.0        2.2.2.2         390         0x80000016      0x0088ab
        10.0.23.0       2.2.2.2         390         0x80000017      0x0023bc
        10.1.3.0        2.2.2.2         390         0x80000018      0x00fcf3
```

由以上输出可知，R2 同时拥有区域 0 和区域 1 的拓扑表，所以 R2 是 BR 和 ABR。R2 在区域 1 中的拓扑表与 R1 的拓扑表完全相同。

```
R3#show ip ospf database
                OSPF Router with ID (3.3.3.3) (Process ID 1)
                    Router Link States (Area 0)       //R3 是骨干路由器（BR）和域内路由器（IR）
        Link ID         ADV Router      Age         Seq#            Checksum Link count
        3.3.3.3         3.3.3.3         462         0x8000000b      0x0096a1  3    //R3 有 3 条链路
        2.2.2.2         2.2.2.2         462         0x8000000c      0x001329  3    //R2 有 3 条链路
                    Summary Net Link States (Area 0)
        Link ID         ADV Router      Age         Seq#            Checksum
        10.0.12.0       2.2.2.2         422         0x8000000f      0x0034fd
        10.1.1.0        2.2.2.2         422         0x80000010      0x00a990
```

由以上输出可知，R3 的拓扑表与路由器 R2 在区域 0 中的拓扑表完全相同。

至此，查看 OSPFv2 拓扑表的实验结束。在本实验中，我们验证了相同 OSPFv2 区域，路由器的拓扑表完全相同。

3.3.4　实验四：查看 OSPFv2 的 LSA

OSPF 路由表是依据链路状态数据库（LSDB），采用 SPF 算法计算生成的。OSPF 网络中有多种路由器角色和区域类型，这些复杂的设计概念要求 OSPF 尽可能交换准确的信息，从而获得最佳路由。为完成这种通信需求，OSPF 使用各种不同类型的 LSA。每种类型的 LSA 负责收集不同的信息，从而形成 LSDB。OSPF 相同区域路由器的 LSDB 相同。1 类 LSA（路由器 LSA）负责收集区域内路由器的链路情况；2 类 LSA（网络 LSA）由 DR 负责收集多路访问网络中所有路由器情况；3 类 LSA（网络汇总 LSA），由 ABR 负责将一个区域内的网络通告给 OSPF 其他区域；5 类 LSA（AS 外部 LSA），由自治系统边界路由器（ASBR）负责将 AS 外部路由注入 OSPF 区域；4 类 LSA（ASBR 汇总 LSA），由 ABR 产生，描述去往 ASBR 的路由。查看 OSPFv2 的 LSA 实验拓扑如图 3-14 所示，下面我们将通过图 3-14 所示拓扑，学习 OSPFv2 LSDB 的 LSA 类型。

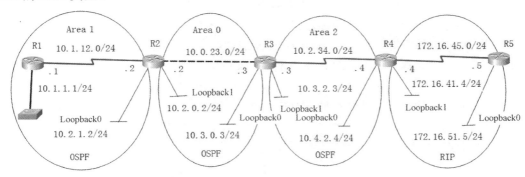

图 3-14　查看 OSPFv2 的 LSA 实验拓扑

（1）查看已经配置好的路由器的路由表

```
R1#show ip route ospf
     10.0.0.0/8 is variably subnetted, 11 subnets, 2 masks
O IA    10.0.23.0/24 [110/65] via 10.1.12.2, 00:13:59, Serial0/0/0      //域间路由（Area 0）
O IA    10.2.0.0/24 [110/65] via 10.1.12.2, 00:13:59, Serial0/0/0       //域间路由（Area 0）
O       10.2.1.0/24 [110/65] via 10.1.12.2, 00:14:44, Serial0/0/0       //域内路由（Area 1）
O IA    10.2.34.0/24 [110/129] via 10.1.12.2, 00:13:59, Serial0/0/0     //域间路由（Area 2）
O IA    10.3.0.0/24 [110/66] via 10.1.12.2, 00:13:59, Serial0/0/0       //域间路由（Area 0）
O IA    10.3.2.0/24 [110/66] via 10.1.12.2, 00:13:59, Serial0/0/0       //域间路由（Area 2）
O IA    10.4.2.0/24 [110/130] via 10.1.12.2, 00:13:59, Serial0/0/0      //域间路由（Area 2）
     172.16.0.0/24 is subnetted, 3 subnets    //OSPFAS 外部路由（RIP 路由域的 3 条路由）
O E2    172.16.41.0/24 [110/20] via 10.1.12.2, 00:14:09, Serial0/0/0    //外部路由（类型 2）
O E2    172.16.45.0/24 [110/20] via 10.1.12.2, 00:14:09, Serial0/0/0    //外部路由（类型 2）
O E2    172.16.51.0/24 [110/20] via 10.1.12.2, 00:14:09, Serial0/0/0    //外部路由（类型 2）
```

由以上输出可知，R1 有 1 条域内路由、6 条域间路由和 3 条 AS 外部路由。

R2#show ip route ospf

```
                10.0.0.0/8 is variably subnetted, 13 subnets, 2 masks
O               10.1.1.0/24 [110/65] via 10.1.12.1, 00:24:17, Serial0/0/0        //域内路由（Area 1）
O IA            10.2.34.0/24 [110/65] via 10.0.23.3, 00:23:47, GigabitEthernet 0/2  //域间路由（Area 2）
O               10.3.0.0/24 [110/2] via 10.0.23.3, 00:23:47, GigabitEthernet 0/2    //域内路由（Area 0）
O IA            10.3.2.0/24 [110/2] via 10.0.23.3, 00:23:47, GigabitEthernet 0/2    //域间路由（Area 2）
O IA            10.4.2.0/24 [110/66] via 10.0.23.3, 00:23:47, GigabitEthernet 0/2   //域间路由（Area 2）
                172.16.0.0/24 is subnetted, 3 subnets //OSPFAS 外部路由              //外部 RIP 路由 3 条
O E2            172.16.41.0/24 [110/20] via 10.0.23.3, 00:23:47, GigabitEthernet 0/2
O E2            172.16.45.0/24 [110/20] via 10.0.23.3, 00:23:47, GigabitEthernet 0/2
O E2            172.16.51.0/24 [110/20] via 10.0.23.3, 00:23:47, GigabitEthernet 0/2
```

由以上输出可知，R2 有 2 条域内路由、3 条域间路由和 3 条 AS 外部路由。我们可以推断路由器 R3 也有 2 条域内路由（10.2.0.0/24 和 10.4.2.0/24）、3 条域间（Area 0 与 Area 1 间）路由和 3 条 AS 外部路由。我们进一步推断路由器 R4 有 1 条域内路由（10.3.2.0/24）、6 条域间（Area 0 与 Area 1 间）路由和 1 条 RIP 协议学习到的路由（172.16.51.0/24）。

接下来，我们查看一下 RIP 路由域中路由器 R5 的路由表。

```
R5>show ip route rip
R       10.0.0.0/8 [120/10] via 172.16.45.4, 00:00:11, Serial0/0/0   //OSPF 域的汇总路由（从 R4 学习到的路由）
        172.16.0.0/16 is variably subnetted, 5 subnets, 2 masks
R       172.16.41.0/24 [120/1] via 172.16.45.4, 00:00:11, Serial0/0/0
```

由以上输出可知，路由器 R5 学习到了由路由器 R4 发布来的度量为 10 的 OSPF 域的汇总路由。

（2）查看路由器 R1 的链路状态数据库（LSDB）

```
R1>show ip ospf database
            OSPF Router with ID (1.1.1.1) (Process ID 1)
                Router Link States (Area 1)       //1 类 LSA，由 Area 1 内路由器生成
Link ID         ADV Router          Age         Seq#            Checksum        Link count
1.1.1.1         1.1.1.1             1303        0x80000006      0x00aaba        3
2.2.2.2         2.2.2.2             1303        0x80000006      0x00312d        3
                Summary Net Link States (Area 1)   //3 类 LSA，由 ABR R2 生成
Link ID         ADV Router          Age         Seq#            Checksum
10.0.23.0       2.2.2.2             1263        0x80000010      0x00b86d
10.2.0.0        2.2.2.2             1263        0x80000011      0x009c9d
10.3.0.0        2.2.2.2             1263        0x80000012      0x00989e
10.3.2.0        2.2.2.2             1263        0x80000013      0x0080b3
10.2.34.0       2.2.2.2             1263        0x80000014      0x00a133
```

Link ID	ADV Router	Age	Seq#	Checksum
10.4.2.0	2.2.2.2	1263	0x80000015	0x00f3fc

Summary ASB Link States (Area 1)   //4 类 LSA，由 ABR R2 生成

Link ID	ADV Router	Age	Seq#	Checksum
4.4.4.4	2.2.2.2	1268	0x8000000f	0x00ed0a

Type-5 AS External Link States   //5 类 LSA，由 ASBR R4 生成，不属于任何区域

Link ID	ADV Router	Age	Seq#	Checksum	Tag
172.16.41.0	4.4.4.4	1311	0x80000003	0x0003c5	0
172.16.45.0	4.4.4.4	1311	0x80000003	0x00d6ed	0
172.16.51.0	4.4.4.4	1304	0x80000003	0x00942a	0

由路由器 R1 的 LSDB 可知，R1 有 1 类、3 类、4 类和 5 类共 4 种 LSA。因为 R1 仅有 Area 1 的 LSDB，所以 R1 的路由器角色为 IR。

（3）查看路由器 R2 的链路状态数据库（LSDB）

```
R2>show ip ospf database
        OSPF Router with ID (2.2.2.2) (Process ID 1)
```

Router Link States (Area 0)   //1 类 LSA，由 Area 0 内路由器生成

Link ID	ADV Router	Age	Seq#	Checksum	Link count
2.2.2.2	2.2.2.2	1456	0x80000005	0x006873	2
3.3.3.3	3.3.3.3	1457	0x80000006	0x003c94	2

Net Link States (Area 0)   //2 类 LSA，由 DR 生成

Link ID	ADV Router	Age	Seq#	Checksum
10.0.23.3	3.3.3.3	1457	0x80000003	0x002de8

Summary Net Link States (Area 0)   //3 类 LSA，由 ABR R2 和 ABR R3 生成

Link ID	ADV Router	Age	Seq#	Checksum
10.2.1.0	2.2.2.2	1446	0x80000007	0x00a59d
10.1.12.0	2.2.2.2	1446	0x80000008	0x00af49
10.1.1.0	2.2.2.2	1446	0x80000009	0x0030d1
10.3.2.0	3.3.3.3	1452	0x8000000a	0x006acf
10.2.34.0	3.3.3.3	1452	0x8000000b	0x008b4f
10.4.2.0	3.3.3.3	1452	0x8000000c	0x00dd19

Summary ASB Link States (Area 0)   //4 类 LSA，由 ABR R3 生成

Link ID	ADV Router	Age	Seq#	Checksum
4.4.4.4	3.3.3.3	1452	0x80000009	0x00db1e

Router Link States (Area 1)   //1 类 LSA，由 Area 1 内路由器生成

Link ID	ADV Router	Age	Seq#	Checksum	Link count
2.2.2.2	2.2.2.2	1486	0x80000006	0x00312d	3
1.1.1.1	1.1.1.1	1487	0x80000006	0x00aaba	3

Summary Net Link States (Area 1)   //3 类 LSA，由 ABR R2 生成

Link ID	ADV Router	Age	Seq#	Checksum
10.0.23.0	2.2.2.2	1446	0x80000010	0x00b86d
10.2.0.0	2.2.2.2	1446	0x80000011	0x009c9d
10.3.0.0	2.2.2.2	1446	0x80000012	0x00989e
10.3.2.0	2.2.2.2	1446	0x80000013	0x0080b3
10.2.34.0	2.2.2.2	1446	0x80000014	0x00a133
10.4.2.0	2.2.2.2	1446	0x80000015	0x00f3fc

Summary ASB Link States (Area 1)   //4 类 LSA，由 ABR R2 生成

Link ID	ADV Router	Age	Seq#	Checksum
4.4.4.4	2.2.2.2	1451	0x8000000f	0x00ed0a

Type-5 AS External Link States   //5 类 LSA，由 ASBR R4 生成，不属于任何区域

Link ID	ADV Router	Age	Seq#	Checksum	Tag
172.16.41.0	4.4.4.4	1495	0x80000003	0x0003c5	0
172.16.45.0	4.4.4.4	1495	0x80000003	0x00d6ed	0
172.16.51.0	4.4.4.4	1488	0x80000003	0x00942a	0

由路由器 R2 的 LSDB 可知，R2 有 1 类、2 类、3 类、4 类和 5 类共 5 种 LSA。因为 R2 同时拥有 Area 0 和 Area 1 的 LSDB，所以 R2 的路由器角色为 BR 和 ABR。

（4）查看路由器 R3 的链路状态数据库（LSDB）

R3>**show ip ospf database**

OSPF Router with ID (3.3.3.3) (Process ID 1)

Router Link States (Area 0)   //1 类 LSA，由 Area 0 内路由器生成

Link ID	ADV Router	Age	Seq#	Checksum	Link count
3.3.3.3	3.3.3.3	1746	0x80000006	0x003c94	2
2.2.2.2	2.2.2.2	1746	0x80000005	0x006873	2

Net Link States (Area 0)   //2 类 LSA，由 DR 生成

Link ID	ADV Router	Age	Seq#	Checksum
10.0.23.3	3.3.3.3	1746	0x80000003	0x002de8

Summary Net Link States (Area 0)   //3 类 LSA，由 ABR R2 和 ABR R3 生成

Link ID	ADV Router	Age	Seq#	Checksum
10.3.2.0	3.3.3.3	1741	0x8000000a	0x006acf
10.2.34.0	3.3.3.3	1741	0x8000000b	0x008b4f
10.4.2.0	3.3.3.3	1741	0x8000000c	0x00dd19
10.2.1.0	2.2.2.2	1736	0x80000007	0x00a59d
10.1.12.0	2.2.2.2	1736	0x80000008	0x00af49
10.1.1.0	2.2.2.2	1736	0x80000009	0x0030d1

Summary ASB Link States (Area 0)   //4 类 LSA，由 ABR R3 生成

Link ID	ADV Router	Age	Seq#	Checksum
4.4.4.4	3.3.3.3	1741	0x80000009	0x00db1e

Router Link States (Area 2)   //1 类 LSA，由 Area 2 内路由器生成

Link ID	ADV Router	Age	Seq#	Checksum	Link count
3.3.3.3	3.3.3.3	1776	0x80000005	0x0042d7	3
4.4.4.4	4.4.4.4	1775	0x80000005	0x00c84a	3

Summary Net Link States (Area 2)   //3 类 LSA，由 ABR R3 生成

Link ID	ADV Router	Age	Seq#	Checksum
10.0.23.0	3.3.3.3	1741	0x8000000d	0x00a084
10.3.0.0	3.3.3.3	1741	0x8000000e	0x0078bf
10.2.0.0	3.3.3.3	1741	0x8000000f	0x008caa
10.2.1.0	3.3.3.3	1731	0x80000010	0x007fb5
10.1.12.0	3.3.3.3	1731	0x80000011	0x008862
10.1.1.0	3.3.3.3	1731	0x80000012	0x000ae9

Type-5 AS External Link States   //5 类 LSA，由 ASBR R4 生成，不属于任何区域

Link ID	ADV Router	Age	Seq#	Checksum	Tag
172.16.41.0	4.4.4.4	1784	0x80000003	0x0003c5	0
172.16.45.0	4.4.4.4	1784	0x80000003	0x00d6ed	0
172.16.51.0	4.4.4.4	1777	0x80000003	0x00942a	0

由路由器 R3 的 LSDB 可知，R3 有 1 类、2 类、3 类、4 类和 5 类共 5 种 LSA。因为 R3 同时拥有 Area 0 和 Area 2 的 LSDB，所以 R3 的路由器角色为 BR 和 ABR。

（5）查看路由器 R4 的链路状态数据库（LSDB）

```
R4>show ip ospf database
```

OSPF Router with ID (4.4.4.4) (Process ID 1)

Router Link States (Area 2)   //1 类 LSA，由 Area 2 内路由器生成

Link ID	ADV Router	Age	Seq#	Checksum	Link count
4.4.4.4	4.4.4.4	620	0x80000006	0x00c64b	3
3.3.3.3	3.3.3.3	623	0x80000006	0x0040d8	3

Summary Net Link States (Area 2)   //3 类 LSA，由 ABR R3 生成

Link ID	ADV Router	Age	Seq#	Checksum
10.0.23.0	3.3.3.3	588	0x80000013	0x00948a
10.3.0.0	3.3.3.3	588	0x80000014	0x006cc5
10.2.0.0	3.3.3.3	588	0x80000015	0x0080b0
10.2.1.0	3.3.3.3	578	0x80000016	0x0073bb
10.1.12.0	3.3.3.3	578	0x80000017	0x007c68

10.1.1.0	3.3.3.3	578	0x80000018	0x00feee	
Type-5 AS External Link States//5 类 LSA，由 ASBR R4 生成，不属于任何区域					
Link ID	ADV Router	Age	Seq#	Checksum	Tag
172.16.41.0	4.4.4.4	629	0x80000004	0x0001c6	0
172.16.45.0	4.4.4.4	629	0x80000004	0x00d4ee	0
172.16.51.0	4.4.4.4	622	0x80000004	0x00922b	0

由路由器 R4 的 LSDB 可知，R4 有 1 类、3 类和 5 类共 3 种 LSA。因为 R4 生成 5 类 LSA，所以，R4 的路由器角色为 ASBR。请记住 5 类 LSA 不属于 OSPF 任何区域。

3.3.5 实验五：验证 OSPFv2 路径

验证 OSPFv2 自动选择最佳路径实验拓扑如图 3-15 所示。

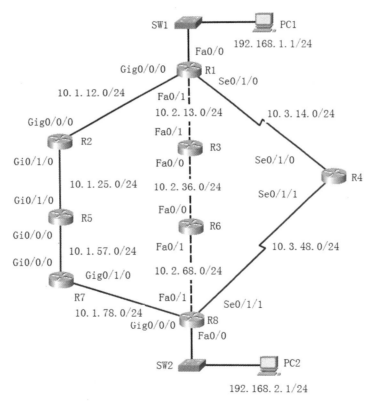

图 3-15 验证 OSPFv2 自动选择最佳路径实验拓扑

在图 3-15 中，8 台路由器都已配置 OSPFv2 并已实现全网互通，请使用 **tracert** 命令验证 PC1 与 PC2 通信数据包经过的路径。

(1) 查看路由器 R1 的路由表

```
R1>show ip route ospf
     10.0.0.0/24 is subnetted, 9 subnets
O       10.1.25.0    [110/2] via 10.1.12.2, 00:18:18, GigabitEthernet 0/0/0
O       10.1.57.0    [110/3] via 10.1.12.2, 00:18:18, GigabitEthernet 0/0/0
O       10.1.78.0    [110/4] via 10.1.12.2, 00:18:18, GigabitEthernet 0/0/0
                     [110/4] via 10.2.13.3, 00:18:18, FastEthernet0/1
O       10.2.36.0    [110/2] via 10.2.13.3, 00:18:18, FastEthernet0/1
O       10.2.68.0    [110/3] via 10.2.13.3, 00:18:18, FastEthernet0/1
O       10.3.48.0    [110/67] via 10.2.13.3, 00:18:18, FastEthernet0/1
O    192.168.2.0/24  [110/4] via 10.2.13.3, 00:18:18, FastEthernet0/1
```

以上 R1 的路由表输出表明,它已添加远程网络的所有路由,去往 PC2 的下一跳为 R3。

(2) 测试 PC1 与 PC2 的连通性

```
C:\>ping 192.168.2.1
Pinging 192.168.2.1 with 32 bytes of data:
Reply from 192.168.2.1: bytes=32 time<1ms TTL=124     //PC1 与 PC2 跨越 4 台路由器(128-4=124)
Reply from 192.168.2.1: bytes=32 time=1ms TTL=124
Reply from 192.168.2.1: bytes=32 time=1ms TTL=124
Reply from 192.168.2.1: bytes=32 time<1ms TTL=124
Ping statistics for 192.168.2.1:
    Packets: Sent = 4, Received = 4, Lost = 0 (0% loss),
Approximate round trip times in milli-seconds:
Minimum = 0ms, Maximum = 1ms, Average = 0ms
```

(3) 验证 PC1 去往 PC2 的路径

```
C:\>tracert 192.168.2.1
Tracing route to 192.168.2.1 over a maximum of 30 hops:
   1    0 ms    0 ms    0 ms    192.168.1.254    //路由器 R1
   2    0 ms    0 ms    0 ms    10.2.13.3        //路由器 R3
   3    0 ms    0 ms    0 ms    10.2.36.6        //路由器 R6
   4    0 ms    0 ms    10 ms   10.2.68.8        //路由器 R8
   5    0 ms    11 ms   0 ms    192.168.2.1      //目的主机 PC2
Trace complete.
```

以上结果表明,主机 PC1 发往 PC2 的数据流经的路径是 PC1→R1→R3→R6→R8→PC2。

（4）验证 PC2 去往 PC1 的路径

```
C:\>tracert 192.168.1.1
Tracing route to 192.168.1.1 over a maximum of 30 hops:
    1    5 ms     0 ms     0 ms    192.168.2.254    //路由器 R8
    2    0 ms     0 ms     0 ms    10.2.68.6        //路由器 R6
    3    0 ms     0 ms     0 ms    10.2.36.3        //路由器 R3
    4   10 ms     0 ms     0 ms    10.2.13.1        //路由器 R1
    5   13 ms    10 ms     0 ms    192.168.1.1      //目的主机 PC1
Trace complete.
```

以上结果表明，主机 PC2 发往 PC1 的数据流经的路径是 PC2→R8→R6→R3→R1→PC1，来回路径一致。显然 OSPFv2 选择的是中间百兆位链路。为什么 OSPFv2 不选择最左面的千兆位链路为最佳路径呢？这个问题我们将在 3.4.4 节场景七中详细介绍。

3.4　干预 OSPFv2 运行

3.4.1　场景四：干预 OSPFv2 邻接关系的建立

经过前面的学习，我们已经了解 OSPFv2 的运行过程。OSPFv2 邻居表、拓扑表以及路由表形成过程是递进的，即先生成邻居表，再产生拓扑表，最后形成路由表。因此，深入研究 OSPFv2 邻接关系的建立，掌握影响邻接关系建立的因素对后期 OSPFv2 网络的维护非常重要。下面我们将在如图 3-16 所示的单区域网络中，在网络运行正常情况下，通过修改不同的参数，干预路由器 R1 和 R2 邻接关系的建立，从而引起网络故障，然后再进行故障恢复。

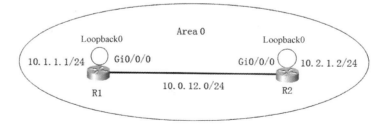

图 3-16　干预 OSPFv2 邻接关系的建立

步骤一：配置单区域 OSPFv2

（1）配置路由器 R1

```
Router(config)#hostname R1
```

```
R1(config)#interface GigabitEthernet 0/0/0
R1(config-if)#ip address 10.0.12.1 255.255.255.0
R1(config-if)#no shutdown
R1(config-if)#interface Loopback 0
R1(config-if)#ip address 10.1.1.1 255.255.255.0
R1(config-if)#router ospf 1
R1(config-router)#router-id 1.1.1.1
R1(config-router)#network 10.0.12.0 0.0.0.255 area 0
R1(config-router)#network 10.1.1.0 0.0.0.255 area 0
```

(2)配置路由器 R2

```
Router(config)#hostname R2
R2(config)#interface GigabitEthernet 0/0/0
R2(config-if)#ip address 10.0.12.2 255.255.255.0
R2(config-if)#no shutdown
R2(config-if)#interface Loopback 0
R2(config-if)#ip address 10.2.1.2 255.255.255.0
R2(config-if)#router ospf 1
R2(config-router)#router-id 2.2.2.2
R2(config-router)#network 10.0.12.0 0.0.0.255 area 0
R2(config-router)#network 10.2.1.0 0.0.0.255 area 0
```

(3)测试网络连通性

```
R1#ping
Protocol [ip]:
Target IP address: 10.2.1.2
Repeat count [5]: 20
Datagram size [100]:
Timeout in seconds [2]:
Extended commands [n]: yes
Source address or interface: 10.1.1.1
Type of service [0]:
Set DF bit in IP header? [no]:
Validate reply data? [no]:
Data pattern [0xABCD]:
Loose, Strict, Record, Timestamp, Verbose[none]:
Sweep range of sizes [n]:
Type escape sequence to abort.
```

```
Sending 20, 100-byte ICMP Echos to 10.2.1.2, timeout is 2 seconds:
Packet sent with a source address of 10.1.1.1
!!!!!!!!!!!!!!!!!!!!
Success rate is 100 percent (20/20), round-trip min/avg/max = 0/0/1 ms
```

由以上输出可知，路由器 R1 采用扩展 **ping** 命令，使用其环回接口（10.1.1.1）ping 通路由器 R2 的环回接口（10.2.1.2），发送了 20 个 ping 包，成功率为 100%。

（4）查看邻接关系

```
R1#show ip ospf neighbor
Neighbor ID     Pri    State          Dead Time      Address        Interface
2.2.2.2         1      FULL/DR        00:00:30       10.0.12.2      GigabitEthernet 0/0/0
```

由以上输出可知，R1 与邻居 R2 已建立完全邻接关系（Full），R2 的角色为 DR。

```
R2#show ip ospf neighbor
Neighbor ID     Pri    State          Dead Time      Address        Interface
1.1.1.1         1      FULL/BDR       00:00:37       10.0.12.1      GigabitEthernet 0/0/0
```

以上输出同样验证了 R2 与 R1 的邻接关系为 Full，R1 为网段 10.0.12.0/24 的 BDR。

步骤二：修改 Hello 和 Dead 间隔干预路由器邻接关系

使用 **show ip ospf interface** 查看 GigabitEthernet0/0/0 接口的 Hello 间隔和 Dead 间隔默认值分别为 10 s 和 40 s。下面我们修改路由器 R1 接口的 Hello 间隔为 5 s，Dead 间隔为 20 s。

```
R1(config)#interface GigabitEthernet 0/0/0
R1(config-if)#ip ospf hello-interval 5
R1(config-if)#ip ospf dead-interval 20
```

接下来，我们查看 R1 邻居表，发现 Dead Time（Dead 间隔）一直递减，至 0 时，邻接状态变为 DOWN（失效状态）。

```
R1#show ip ospf neighbor
Neighbor ID     Pri    State          Dead Time      Address        Interface
2.2.2.2         1      FULL/DR        00:00:00       10.0.12.2      GigabitEthernet 0/0/0
16:33:45: %OSPF-5-ADJCHG: Process 1, Nbr 2.2.2.2 on GigabitEthernet 0/0/0 from FULL to DOWN,
Neighbor Down: Dead timer expired
16:33:45: %OSPF-5-ADJCHG: Process 1, Nbr 2.2.2.2 on GigabitEthernet 0/0/0 from FULL to DOWN,
Neighbor Down: Interface down or detached
R1#show ip ospf neighbor

R1#
```

以上实验证明，链路两邻居的 Hello 间隔和 Dead 间隔不一致会影响邻接建立。
请修改路由器 R2 的对应参数或还原 R1 默认设置，排除网络故障。

步骤三：修改接口优先级干预路由器 OSPF 邻接关系

请先确保网络故障已被排除，R1 和 R2 邻接关系正常，然后再进行以下操作。请将 R2 接口优先级修改为 0，重置 R2 的 OSPFv2 进程。

```
R2(config)#interface GigabitEthernet 0/0/0
R2(config-if)#ip ospf priority 0
R2(config-if)#do clear ip ospf process
Reset ALL OSPF processes? [no]: yes
#
```

请将 R1 接口优先级修改为 0，重置 OSPFv2 进程，重新建立邻接关系。

```
R1(config)#interface GigabitEthernet 0/0/0
R1(config-if)#ip ospf priority 0
R1(config-if)#end
R1#clear ip ospf process
Reset ALL OSPF processes? [no]: yes
17:10:03: %OSPF-5-ADJCHG: Process 1, Nbr 2.2.2.2 on GigabitEthernet 0/0/0 from FULL to DOWN,Neighbor Down: Adjacency forced to reset
17:10:03: %OSPF-5-ADJCHG: Process 1, Nbr 2.2.2.2 on GigabitEthernet 0/0/0 from FULL to DOWN, Neighbor Down: Interface down or detached
R1#show ip ospf neighbor
Neighbor ID     Pri    State            Dead Time   Address      Interface
2.2.2.2         0      2WAY/DROTHER     00:00:33    10.0.12.2    GigabitEthernet 0/0/0
```

以上实验证明，修改接口优先级后重置 OSPFv2 进程，R1 和 R2 的邻接关系从 **FULL** 状态变成 **2WAY** 状态，网络中不存在 DR 和 BDR，R1 和 R2 角色均变成 DROTHER。以上邻接关系出现问题，引起拓扑表无法建立，进而路由表无法形成。

在广播式多路访问类型网络中，我们修改路由器接口优先级为 0，意味着剥夺 DR 选举权，因此，可以将至少一台路由器接口优先级改为非 0，由此产生 DR 并排除网络故障。

当然，我们也可以通过修改接口网络类型的方法，将 BROADCAST（广播类型）更改为 POINT-TO-POINT（点到点类型），也可以排除当前网络故障。下面我们通过实验验证一下：

```
R2(config)#interface GigabitEthernet 0/0/0
R2(config-if)#ip ospf network point-to-point
17:40:31: %OSPF-5-ADJCHG: Process 1, Nbr 1.1.1.1 on GigabitEthernet 0/0/0 from 2WAY to DOWN,Neighbor Down: Interface down or detached
```

由以上可知，修改 R2 接口网络类型为点到点后，R2 与 R1 的邻接关系从 **2WAY** 状态变成 **DOWN** 状态。

  R1(config)#**interface GigabitEthernet 0/0/0**
  R1(config-if)#**ip ospf network point-to-point**
  17:42:27: %OSPF-5-ADJCHG: Process 1, Nbr 2.2.2.2 on GigabitEthernet 0/0/0 from **2WAY to DOWN,Neighbor Down**: Interface down or detached
  17:42:27: %OSPF-5-ADJCHG: Process 1, Nbr 2.2.2.2 on GigabitEthernet 0/0/0 from **LOADING to FULL,** Loading Done

由以上可知，修改 R1 接口网络类型为点到点后，R2 与 R1 邻接关系先从 **2WAY** 状态变成 **DOWN** 状态，最终回到 **FULL** 状态。

  R1#**show ip ospf neighbor**
  Neighbor ID Pri State   Dead Time Address   Interface
  2.2.2.2   0   FULL/ -   00:00:35   10.0.12.2 GigabitEthernet 0/0/0

以上实验证明，修改网络接口类型为点到点，邻接状态变为"FULL/ -"，即不选举 DR，也可排除网络故障。

  R1# **ping 10.2.1.2**
  Sending 5, 100-byte ICMP Echos to 10.2.1.2, timeout is 2 seconds:
  !!!!!
  Success rate is 100 percent (5/5), round-trip min/avg/max = 0/0/1 ms

此时，路由器 R1 可以正常访问 R2 的环回接口，实现网络互通。

步骤四：修改其他参数值干预路由器 OSPF 邻接关系

（1）修改接口子网掩码干预邻接关系

请先确保网络故障已被排除，R1 和 R2 邻接关系正常，然后再进行以下操作。

  R1(config)#**interface GigabitEthernet 0/0/0**
  R1(config-if)#**ip address 10.0.12.1 255.255.255.128**
  18:18:28: %OSPF-5-ADJCHG: Process 1, Nbr 2.2.2.2 on GigabitEthernet 0/0/0 **from FULL to DOWN,** Neighbor Down: Interface down or detached

由上可知，将 R1 接口子网掩码改为/25 后，R1 和 R2 邻接关系从 **FULL** 状态变成 **DOWN** 状态。

  R1(config-if)#**ip address 10.0.12.1 255.255.255.0**
  18:18:47: %OSPF-5-ADJCHG: Process 1, Nbr 2.2.2.2 on GigabitEthernet 0/0/0 **from LOADING to FULL,** Loading Done

由以上可知，再将 R1 接口子网掩码改回（同 R2 接口子网掩码一样），R1 与 R2 邻接关系回到 **FULL** 状态。

（2）修改接口区域号干预邻接关系

请先确保网络故障已被排除，R1 和 R2 邻接关系正常，然后再进行以下操作。

R1(config)#**router ospf 1**
R1(config-router)#**network 10.0.12.0 0.0.0.255 area 1**
18:22:11: %OSPF-5-ADJCHG: Process 1, Nbr 2.2.2.2 on GigabitEthernet 0/0/0 from **FULL to DOWN**, Neighbor Down: Interface down or detached
18:22:11: %OSPF-6-AREACHG: 10.0.12.0/0 changed **from area 0 to area 1**
18:22:13: %OSPF-4-ERRRCV: Received invalid packet: **mismatch area ID,** from backbone area must be virtual-link but not found from 10.0.12.1, GigabitEthernet 0/0/0

由以上可知，修改路由器 R1 接口区域号使其与路由器 R2 接口的区域号不一致，邻接关系变为 **DOWN** 状态。

R1(config-router)#**network 10.0.12.0 0.0.0.255 area 0**
18:22:23: %OSPF-6-AREACHG: 10.0.12.0/0 **changed from area 1 to area 0**
18:22:23: %OSPF-5-ADJCHG: Process 1, Nbr 2.2.2.2 on GigabitEthernet 0/0/0 **from LOADING to FULL, Loading Done**

由以上可知，修改路由器 R1 接口区域号使其与路由器 R2 接口区域号相同，邻接关系回到 **FULL** 状态。

（3）修改接口路由器 ID 干预邻接关系

请先确保网络故障已被排除，R1 和 R2 邻接关系正常，然后再进行以下操作。

R1(config)#**router ospf 1**
R1(config-router)#**router-id 2.2.2.2**
R1(config-router)#**do clear ip ospf process**
Reset ALL OSPF processes? [no]: **yes**
18:29:13: %OSPF-5-ADJCHG: Process 1, Nbr 2.2.2.2 on GigabitEthernet 0/0/0 from **FULL to DOWN**, Neighbor Down: **Adjacency forced to reset**
18:29:13: %OSPF-5-ADJCHG: Process 1, Nbr 2.2.2.2 on GigabitEthernet 0/0/0 from **FULL to DOWN**, Neighbor Down: **Interface down or detached**

以上实验证明，修改 R1 的路由器 ID 使其与 R2 的路由器 ID 相同，在清除 OSPFv2 进程后，R1 与 R2 邻接关系从 FULL 状态变成 DOWN 状态。我们再次查看 R1 的邻居表：

R1#**show ip ospf neighbor**

Neighbor ID	Pri	State	Dead Time	Address	Interface
2.2.2.2	1	**EXSTART/BDR**	00:00:34	10.0.12.2	GigabitEthernet 0/0/0

以上输出表明，R1 与 R2 邻接状态变成了 **EXSTART**（不稳定邻接）状态，由于两台路由器的路由器 ID 冲突，双方无法交换 DBD 数据包，也就无法建立 LSDB。接下来，我们将 R1

的路由器 ID 改回。

```
R1(config-router)#router-id 1.1.1.1
Reload or use "clear ip ospf process" command, for this to take effect
R1(config-router)#end
R1#clear ip ospf process
Reset ALL OSPF processes? [no]: yes
18:30:55: %OSPF-5-ADJCHG: Process 1, Nbr 2.2.2.2 on GigabitEthernet 0/0/0 from EXSTART to DOWN, Neighbor Down: Adjacency forced to reset
18:30:55: %OSPF-5-ADJCHG: Process 1, Nbr 2.2.2.2 on GigabitEthernet 0/0/0 from EXSTART to DOWN, Neighbor Down: Interface down or detached
18:31:03: %OSPF-5-ADJCHG: Process 1, Nbr 2.2.2.2 on GigabitEthernet 0/0/0 from LOADING to FULL, Loading Done
```

以上输出表明，解决路由器 ID 冲突后问题，故障便被排除。本场景实验证明，Hello 间隔、Dead 间隔、子网掩码和区域号等不匹配，以及路由器 ID 冲突都会影响 OSPFv2 邻接关系的建立。当然影响邻接关系建立的因素远不止这些，例如，认证类型、密钥和特殊区域等不匹配也会影响邻接关系。

3.4.2　场景五：干预 OSPFv2 DR 选举

某企业网络运行性能较差，经排查，发现原因是由一台性能最低的路由器 R5 担任企业网 DR。现在要让性能最高的路由器 R1 成为 DR，R5 不能成为 DR 和 BDR，我们该如何实现？接下来，让我们一起通过实验干预 OSPFv2 DR 选举，实验拓扑如图 3-17 所示。

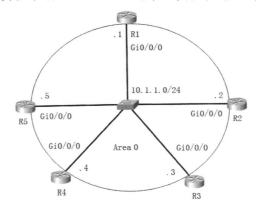

图 3-17　干预 OSPFv2 DR 选举实验拓扑

方法一　更改路由器接口优先级干预 DR 选举

（1）查看路由器邻居表了解路由器角色

```
R1#show ip ospf neighbor
```

Neighbor ID	Pri	State	Dead Time	Address	Interface
10.1.1.4	1	**FULL/BDR**	00:00:35	10.1.1.4	GigabitEthernet 0/0/0
10.1.1.5	1	**FULL/DR**	00:00:35	10.1.1.5	GigabitEthernet 0/0/0
10.1.1.3	1	2WAY/DROTHER	00:00:35	10.1.1.3	GigabitEthernet 0/0/0
10.1.1.2	1	2WAY/DROTHER	00:00:35	10.1.1.2	GigabitEthernet 0/0/0

由以上输出可知，路由器 R5 担任网段 10.1.1.0/24 的 DR，R4 是 BDR。

（2）修改路由器接口优先级干预 DR 选举

```
R1(config)#interface GigabitEthernet 0/0/0
R1(config-if)#ip ospf priority 255
R1(config-if)#end
R1#clear ip ospf process
Reset ALL OSPF processes? [no]: yes
```

由上 R1 的配置可知，已将其接口优先级设置为最大值 255，其目的是确保 R1 一定有 DR 选举权。

```
R5(config)#interface GigabitEthernet 0/0/0
R5(config-if)#ip ospf priority 0
R5(config-if)#end
R5#clear ip ospf process
Reset ALL OSPF processes? [no]: yes
```

由以上 R5 的配置可知，已将其接口优先级设置为最小值 0，其目的是剥夺 R5 的 DR 选举权。

（3）验证当前路由器 R1 和 R5 的角色

在路由器 R2 上查看邻居 R1 和 R5 的路由器角色。

```
R2>show ip ospf neighbor
```

Neighbor ID	Pri	State	Dead Time	Address	Interface
10.1.1.3	1	2WAY/DROTHER	00:00:33	10.1.1.3	GigabitEthernet 0/0/0
10.1.1.1	255	**FULL/DR**	00:00:33	10.1.1.1	GigabitEthernet 0/0/0
10.1.1.4	1	FULL/BDR	00:00:33	10.1.1.4	GigabitEthernet 0/0/0
10.1.1.5	0	2WAY/DROTHER	00:00:33	10.1.1.5	GigabitEthernet 0/0/0

以上输出结果可知，路由器 R1 已经成功当选网段 DR，R5 的角色为网段的 DROTHER。

方法二：更改路由器 ID 干预 DR 选举

（1）查看路由器邻居表了解路由器角色

```
R1#show ip ospf neighbor
```

Neighbor ID	Pri	State	Dead Time	Address	Interface
10.1.1.5	1	**FULL/DR**	00:00:35	10.1.1.5	GigabitEthernet 0/0/0
10.1.1.4	1	**FULL/BDR**	00:00:35	10.1.1.4	GigabitEthernet 0/0/0
10.1.1.2	1	2WAY/DROTHER	00:00:35	10.1.1.2	GigabitEthernet 0/0/0
10.1.1.3	1	2WAY/DROTHER	00:00:35	10.1.1.3	GigabitEthernet 0/0/0

由以上输出可知，路由器 R5 担任网段 10.1.1.0/24 的 DR，R4 是 BDR。

（2）修改路由器 ID 以干预 DR 选举

修改 R1 的路由器 ID 为 221.6.20.1，此时 R1 路由器 ID 在本网段最大。

```
R1(config)#router ospf 1
R1(config-router)#router-id 221.6.20.1
R1(config-router)#end
R1#clear ip ospf process
Reset ALL OSPF processes? [no]: y
```

修改 R5 的路由器 ID 为 5.5.5.5，此时 R5 的路由器 ID 在本网段最小。

```
R5(config)#router ospf 1
R5(config-router)#router-id 5.5.5.5
R5(config-router)#end
R5#clear ip ospf process
Reset ALL OSPF processes? [no]: y
```

（3）验证当前路由器 R1 和 R5 的角色

在路由器 R2 上查看邻居 R1 和 R5 的路由器角色。

```
R2>show ip ospf neighbor
```

Neighbor ID	Pri	State	Dead Time	Address	Interface
10.1.1.3	1	2WAY/DROTHER	00:00:32	10.1.1.3	GigabitEthernet 0/0/0
221.6.20.1	1	**FULL/DR**	00:00:32	10.1.1.1	GigabitEthernet 0/0/0
10.1.1.4	1	FULL/BDR	00:00:32	10.1.1.4	GigabitEthernet 0/0/0
5.5.5.5	1	**2WAY/DROTHER**	00:00:32	10.1.1.5	GigabitEthernet 0/0/0

由以上输出结果可知，路由器 R1 已经成功当选网段 DR，R5 成为 DROTHER。

至此，干预 OSPFv2 DR 选举实验结束。在本实验中，我们采用了修改路由器接口优先级和修改路由器 ID 两种方法成功干预 DR 选举。请读者注意修改完相应参数后一定要重置 OSPFv2 进程。我们知道，在多路访问型网络中要进行 DR 选举，其选举原则是先查看接口优先级，优先级最高的路由器当选；在优先级相同的情况下，路由器 ID 最大的路由器当选。

3.4.3 场景六：优化 OSPFv2 网络性能

如图 3-18 所示，要求 3 台路由器运行 OSPFv2。我们知道该 OSPFv2 是路由器之间的通信协议。默认情况下，路由器会向其所有接口发送消息。实际上，发送给交换机泛洪至终端没有意义，浪费链路、CPU 及内存资源。因此我们可将连接业务网段的路由器接口设置成被动接口，以提升网络性能。路由器 R1 和 R2 之间链路是广播型多路访型链路，链路连接两台设备，DR/BDR 选举也没有必要。下面让我们一起配置 OSPFv2 网络，并对其性能进行优化。

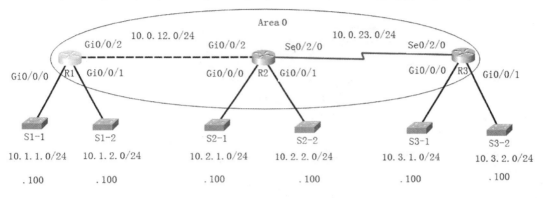

图 3-18　优化 OSPFv2 网络性能

步骤一：配置单区域 OSPFv2 实现全网互通

（1）在路由器 R1 上配置接口地址和 OSPFv2

```
R1(config)#interface GigabitEthernet 0/0/0
R1(config-if)#ip add 10.1.1.254 255.255.255.0
R1(config-if)#no shutdown
R1(config-if)#interface GigabitEthernet 0/0/1
R1(config-if)#ip address 10.1.2.254 255.255.255.0
R1(config-if)#no shutdown
R1(config-if)#interface GigabitEthernet 0/0/2
R1(config-if)#ip address 10.0.12.1 255.255.255.0
R1(config-if)#no shutdown
R1(config-if)#router ospf 1
R1(config-router)#router-id 1.1.1.1
R1(config-router)#network 10.1.1.0 0.0.0.255 area 0
R1(config-router)#network 10.1.2.0 0.0.0.255 area 0
R1(config-router)#network 10.0.12.0 0.0.0.255 area 0
```

（2）在路由器 R2 上配置接口地址和 OSPFv2

```
R2(config)#interface GigabitEthernet 0/0/0
R2(config-if)#ip address 10.2.1.254 255.255.255.0
R2(config-if)#no shutdown
R2(config-if)#interface GigabitEthernet 0/0/1
R2(config-if)#ip address 10.2.2.254 255.255.255.0
R2(config-if)#no shutdown
R2(config-if)#interface GigabitEthernet 0/0/2
R2(config-if)#ip address 10.0.12.2 255.255.255.0
R2(config-if)#no shutdown
R2(config-if)#interface serial0/2/0
R2(config-if)#ip address 10.0.23.2 255.255.255.0
R2(config-if)#no shutdown
R2(config-if)#router ospf 1
R2(config-router)#router-id 2.2.2.2
R2(config-router)#network 10.2.1.0 0.0.0.255 area 0
R2(config-router)#network 10.2.2.0 0.0.0.255 area 0
R2(config-router)#network 10.0.12.0 0.0.0.255 area 0
R2(config-router)#network 10.0.23.0 0.0.0.255 area 0
```

（3）在路由器 R3 上配置接口地址和 OSPFv2

```
R3(config)#interface GigabitEthernet 0/0/0
R3(config-if)#ip address 10.3.1.254 255.255.255.0
R3(config-if)#no shutdown
R3(config-if)#interface GigabitEthernet 0/0/1
R3(config-if)#ip address 10.3.2.254 255.255.255.0
R3(config-if)#no shutdown
R3(config-if)#interface serial0/2/0
R3(config-if)#ip address 10.0.23.3 255.255.255.0
R3(config-if)#no shutdown
R3(config-if)#router ospf 1
R3(config-router)#router-id 3.3.3.3
R3(config-router)#network 10.3.1.0 0.0.0.255 area 0
R3(config-router)#network 10.3.2.0 0.0.0.255 area 0
R3(config-router)#network 10.0.23.0 0.0.0.255 area 0
```

（4）在交换机上配置管理 IP 地址和默认网关

> S1-1(config)#**interface vlan 1**
> S1-1(config-if)#**ip address 10.1.1.100 255.255.255.0**
> S1-1(config-if)#**no shutdown**
> S1-1(config-if)#**exit**
> S1-1(config)#**ip default-Gateway 10.1.1.254**

请参照交换机 S1-1 的配置，配置其余 5 台交换机（S1-2、S2-1、S2-2、S3-1 和 S3-2）的 IP 地址和默认网关。

（5）网络连通性测试

如下所示，使用交换机作为终端依次对不同网段进行连通性测试。

> S1-1#**ping 10.3.2.100**
> Sending 5, 100-byte ICMP Echos to 10.3.2.100, timeout is 2 seconds:
> !!!!!
> Success rate is 100 percent (5/5), round-trip min/avg/max = 1/4/17 ms

此时，各网段间可以实现互访，全网互通。

步骤二：配置被动接口，优化网络性能

（1）在路由器 R1 上配置被动接口

> R1(config)#**router ospf 1**
> R1(config-router)# **passive-interface GigabitEthernet 0/0/0**
> R1(config-router)# **passive-interface GigabitEthernet 0/0/1**

（2）在路由器 R2 上配置被动接口

> R2(config)#**router ospf 1**
> R2(config-router)# **passive-interface GigabitEthernet 0/0/0**
> R2(config-router)# **passive-interface GigabitEthernet 0/0/1**

（3）在路由器 R3 上配置被动接口

> R3(config)#**router ospf 1**
> R3(config-router)# **passive-interface GigabitEthernet 0/0/0**
> R3(config-router)# **passive-interface GigabitEthernet 0/0/1**

（4）查看配置的被动接口

以下输出表明，在路由器 R1 上已经设置了被动接口，OSPFv2 消息将不会从相应接口转发。

> R1#**show ip protocols**

```
Routing Protocol is "ospf 1"
<Output omitted>
   Passive Interface(s):
      GigabitEthernet 0/0/0
      GigabitEthernet 0/0/1
<Output omitted>
```

步骤三：更改网络类型，优化网络性能

（1）查看路由器 R1 和 R2 的邻居表

```
R1>show ip ospf neighbor
Neighbor ID     Pri    State      De Time     Address       Interface
2.2.2.2          1     FULL/DR    00:00:34    10.0.12.2     GigabitEthernet 0/0/2
```

以上输出表明，在广播多路访问型网络 10.0.12.0/24 中，路由器 R2 当选 DR。

```
R2#show ip ospf neighbor
Neighbor ID     Pri    State      Dead Time   Address       Interface
3.3.3.3          0     FULL/  -   00:00:37    10.0.23.3     Serial0/2/0
1.1.1.1          1     FULL/BDR   00:00:37    10.0.12.1     GigabitEthernet 0/0/2
```

以上输出表明，在网段 10.0.12.0/24 中，路由器 R1 当选 BDR。

（2）更改路由器接口网络类型为点到点

① 将路由器 R1 的接口网络类型改成点到点。

```
R1(config)#interface GigabitEthernet 0/0/2
R1(config-if)#ip ospf network point-to-point
08:31:04: %OSPF-5-ADJCHG: Process 1, Nbr 2.2.2.2 on GigabitEthernet 0/0/2 from FULL to DOWN, Neighbor Down: Interface down or detached
08:31:04: %OSPF-5-ADJCHG: Process 1, Nbr 2.2.2.2 on GigabitEthernet 0/0/2 from LOADING to FULL, Loading Done
```

② 将路由器 R2 接口网络类型改成点到点。

```
R2(config)#interface GigabitEthernet 0/0/2
R2(config-if)#ip ospf network point-to-point
08:33:12: %OSPF-5-ADJCHG: Process 1, Nbr 1.1.1.1 on GigabitEthernet 0/0/2 from FULL to DOWN, Neighbor Down: Interface down or detached
08:33:12: %OSPF-5-ADJCHG: Process 1, Nbr 1.1.1.1 on GigabitEthernet 0/0/2 from LOADING to FULL, Loading Done
```

以上日志消息表明，修改 R1 和 R2 网络类型后，邻接关系重新建立。此时邻接表如下：

```
R1#show ip ospf neighbor
Neighbor ID     Pri     State       Dead Time       Address         Interface
2.2.2.2         0       FULL/ -     00:00:37        10.0.12.2       GigabitEthernet 0/0/2
```

以上输出表明，此时网段不选举 DR，接口优先级为 0，R1 和 R2 完全邻接。

```
R2#show ip ospf neighbor
Neighbor ID     Pri     State       Dead Time       Address         Interface
3.3.3.3         0       FULL/ -     00:00:31        10.0.23.3       Serial0/2/0
1.1.1.1         0       FULL/ -     00:00:35        10.0.12.1       GigabitEthernet 0/0/2
```

以上输出表明，分别用串行口和以太网口连接的 WAN 和 LAN 都不进行 DR 选举，因为网络类型都是点到点，这样可以提升网络收敛时间，优化网络性能。

```
R1>show ip route ospf
     10.0.0.0/8 is variably subnetted, 11 subnets, 2 masks
O       10.0.23.0 [110/65] via 10.0.12.2, 00:01:21, GigabitEthernet 0/0/2
O       10.2.1.0 [110/2] via 10.0.12.2, 00:01:21, GigabitEthernet 0/0/2
O       10.2.2.0 [110/2] via 10.0.12.2, 00:01:21, GigabitEthernet 0/0/2
O       10.3.1.0 [110/66] via 10.0.12.2, 00:01:11, GigabitEthernet 0/0/2
O       10.3.2.0 [110/66] via 10.0.12.2, 00:01:11, GigabitEthernet 0/0/2
```

以上输出表明，修改接口网络类型后，路由表完全正常，网间通信正常。

至此，优化 OSPFv2 网络性能的实验结束。在本实验中，我们通过设置被动接口和修改网络接口类型达到了优化网络性能目的。第 4 章我们还会学习通过 OSPFv2 路由汇总和在特殊区域优化路由表，提升网络性能。

3.4.4　场景七：干预 OSPFv2 自动选路

通过 3.3.5 节实验五我们知道，OSPFv2 选择了中间百兆位链路而没有选择千兆位链路作为最佳路径，什么原因呢？OSPF 使用开销（Cost）作为度量标准，开销指从路由器到目的网络的累计开销，开销越小，路径越优。开销与接口带宽成反比。带宽越宽，开销越小（Cost=参考带宽／接口带宽），默认参考带宽为 10^8 bps。由于 OSPF 开销值必须是正整数，因此百兆位和千兆位接口开销相同，都为 1，这样千兆位链路累计开销为 5，百兆位链路累计开销为 4。要实现精确选路，可用 **auto-cost reference-bandwidth** 命令调整参考带宽或使用 **ip ospf cost** 命令手动设置开销值。干预 OSPFv2 选路实验拓扑如图 3-19 所示。

步骤一：查看 PC1 通往 PC2 的默认路径

（1）查看路由器 R1 的路由表了解下一跳

```
R1>show ip route | include 192.168.2.0
O    192.168.2.0/24 [110/4] via 10.2.13.3, 00:18:18, FastEthernet0/1
```

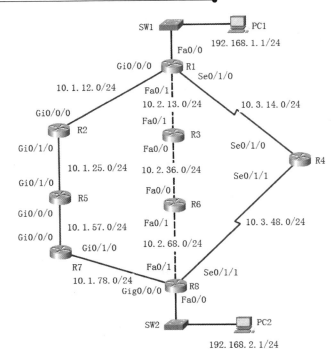

图 3-19 干预 OSPFv2 选路实验拓扑

以上输出表明，通往 PC2 所在网段路径开销为 4，下一跳为 R3，送出接口为 Fa0/1。

（2）查看 PC1 通往 PC2 数据流经路径

```
C:\>tracert 192.168.2.1
Tracing route to 192.168.2.1 over a maximum of 30 hops:
  1    0 ms      0 ms      0 ms     192.168.1.254    //R1
  2    0 ms      0 ms      0 ms     10.2.13.3        //R3
  3    0 ms      0 ms      0 ms     10.2.36.6        //R6
  4    0 ms      0 ms     10 ms     10.2.68.8        //R8
  5   11 ms      0 ms      0 ms     192.168.2.1      //目的主机 PC2
Trace complete.
```

以上输出表明，主机 PC1 发往 PC2 的数据流经路径是 PC1→R1→R3→R6→R8→PC2。

步骤二：查看 R1 通往 PC2 的路径（模拟故障，切换至备份路径）

（1）使路由器 R1 的百兆位接口失效，查看 R1 的路由表

```
R1(config)#interface FastEthernet 0/1
R1(config-if)#shutdown
```

```
R1>show ip route | include 192.168.2.0
O        192.168.2.0/24 [110/5] via 10.1.12.2, 00:23:46, GigabitEthernet 0/0/0
```

以上输出表明，通往主机 PC2 所在网段开销变为 5，下一跳为 R2，送出接口切换至 Gi0/0/0。

```
C:\>tracert 192.168.2.1
Tracing route to 192.168.2.1 over a maximum of 30 hops:
  1    0 ms    0 ms    1 ms    192.168.1.254    //R1
  2    0 ms    0 ms    0 ms    10.1.12.2        //R2
  3    0 ms    0 ms    0 ms    10.1.25.5        //R5
  4   10 ms    0 ms    0 ms    10.1.57.7        //R7
  5   11 ms   10 ms    0 ms    10.1.78.8        //R8
  6   13 ms   10 ms    0 ms    192.168.2.1      //目的主机 PC2
Trace complete.
```

以上输出表明，PC1 发往 PC2 的数据路径切换至 PC1→R1→R2→R5→R7→R8→PC2。

（2）继续使路由器 R1 的千兆位接口失效，查看 R1 的路由表

```
R1(config)#interface GigabitEhternet 0/0/0
R1(config-if)#shutdown
R1>show ip route | begin 192.168.2.0
O        192.168.2.0/24 [110/129] via 10.3.14.4, 00:00:38, Serial0/1/0
```

以上输出表明，通往主机 PC2 路径开销变为 129，下一跳为 R4，送出接口为 Se0/1/0。

```
C:\>tracert 192.168.2.1
Tracing route to 192.168.2.1 over a maximum of 30 hops:
  1    0 ms    0 ms    0 ms    192.168.1.254    //R1
  2    0 ms    0 ms    0 ms    10.3.14.4        //R4
  3   17 ms    2 ms   12 ms    10.3.48.8        //R8
  4    0 ms    1 ms    0 ms    192.168.2.1      //目的主机 PC2
Trace complete.
```

以上输出表明，PC1 发往 PC2 的数据路径切换到 PC1→R1→R4→R8→PC2。

步骤三：干预 OSPFv2 选路，选择最佳路径（修改参考带宽方法）

（1）修改路由器 R1 至 R8 的参考带宽为 10^9 bps（1000 Mbps）

```
R1(config-router)#auto-cost reference-bandwidth ?
  <1-4294967>   The reference bandwidth in terms of Mbits per second    //参考带宽单位为 Mbps
R1(config-router)#auto-cost reference-bandwidth 1000
% OSPF: Reference bandwidth is changed.                                 //参考带宽被修改
     Please ensure reference bandwidth is consistent across all routers.
```

请读者参考 R1，完成其余路由器参考带宽值的修改，确保所有路由器的参考带宽一致。

（2）查看路由器 R1 通往主机 PC2 的路径开销

```
R1#show ip route ospf
     10.0.0.0/24 is subnetted, 9 subnets
O       10.1.25.0 [110/2] via 10.1.12.2, 00:07:58, GigabitEthernet 0/0/0
O       10.1.57.0 [110/3] via 10.1.12.2, 00:07:58, GigabitEthernet 0/0/0
O       10.1.78.0 [110/4] via 10.1.12.2, 00:01:27, GigabitEthernet 0/0/0
O       10.2.36.0 [110/20] via 10.2.13.3, 00:00:47, FastEthernet0/1
O       10.2.68.0 [110/14] via 10.1.12.2, 00:00:47, GigabitEthernet 0/0/0
O       10.3.48.0 [110/651] via 10.1.12.2, 00:00:47, GigabitEthernet 0/0/0
O    192.168.2.0/24 [110/14] via 10.1.12.2, 00:00:47, GigabitEthernet 0/0/0   //1+1+1+1+10=14
```

以上输出表明，R1 通往主机 PC2 的路径开销为 14，从 R2 千兆位接口转发，干预选路成功。我们使用 **show ip ospf interface** *interface* 命令可以查看千兆位接口开销为 1，百兆位接口开销为 10，因此 OSPFv2 可以准确选出最佳路径。

（3）查看主机 PC1 通往目的主机 PC2 的路径

```
C:\>tracert 192.168.2.1
Tracing route to 192.168.2.1 over a maximum of 30 hops:
  1    0 ms      0 ms     0 ms      192.168.1.254    //R1
  2    0 ms      1 ms     0 ms      10.1.12.2        //R2
  3    0 ms      0 ms     0 ms      10.1.25.5        //R5
  4   11 ms      0 ms    11 ms      10.1.57.7        //R7
  5    0 ms     13 ms    12 ms      10.1.78.8        //R8
  6   15 ms     22 ms    10 ms      192.168.2.1      //目的主机 PC2
Trace complete.
```

以上输出表明，PC1 发往 PC2 的路径为理想的千兆位宽带宽路径，干预选路成功。

步骤四：干预 OSPFv2 选路，选择最佳路径（修改接口开销方法）

默认 PC1 发往 PC2 的数据流经百兆位链路 R1→R3→R6→R8，路径累计开销为 4，如下所示。

```
R1#show ip route | include 192.168.2.0
O    192.168.2.0/24 [110/4] via 10.2.13.3, 00:00:45, FastEthernet0/1   //开销为4，下一跳为R3
```

因为千兆位路径 R1→R2→R5→R7→R8 累计开销为 5，所以默认下一跳 R2 不会写进路由表。

接下来，我们修改 R1 接口开销，以达到使百兆位链路累计开销大于千兆位链路累计开销的目的。

```
R1(config)#interface FastEthernet 0/1
R1(config-if)#ip ospf cost 3
```

将 R1 百兆位接口开销改为 3，则百兆位链路累计开销 6 大于千兆位链路累计开销。我们查看路由条目的变化，如下所示。

```
R1#show ip route | include 192.168.2.0
O    192.168.2.0/24 [110/5] via 10.1.12.2, 00:00:08, GigabitEthernet 0/0/0
```

以上输出表明，通往 PC2 路径开销为 5，下一跳为 R2，采用千兆位接口转发，干预选路成功。

为确保 PC1 与 PC2 通信来回路径一致，我们查看路由器 R8 去往 PC1 的下一跳。

```
R8#show ip route | include 192.168.1.0
O    192.168.1.0/24 [110/4] via 10.2.68.6, 00:11:15, FastEthernet0/1    //开销为4,下一跳路由器为R6
```

以上输出表明，通往 PC1 路径开销为 4，下一跳为 R6，送出接口为百兆接口。

```
R8(config)#interface FastEthernet 0/1
R8(config-if)#ip ospf cost 3
```

以上操作修改 R8 的百兆位接口开销为 3，这样百兆位链路开销累计为 6，此时路由条目变化如下。

```
R8#show ip route | include 192.168.1.0
O    192.168.1.0/24 [110/5] via 10.1.78.7, 00:00:03, GigabitEthernet 0/0/0
```

以上输出表明，到 PC1 路径开销为 5，下一跳为 R7，采用千兆位接口转发，干预选路成功。

至此，干预 OSPFv2 自动选路实验结束。本实验采用修改参考带宽和修改接口开销两种方法成功干预选路。掌握 OSPFv2 路径选择标准，可以干预选路，从而提升网络性能。

3.5 挑战练习

3.5.1 挑战练习一

挑战要求：图 3-20 所示网络拓扑由 5 台路由器和 5 台交换机互联而成。设备间连接接口及网络地址分配见图中标注。请根据本章所学知识完成 OSPFv2 相关配置，确保全网互通。

请读者根据任务要求独立完成该挑战练习。

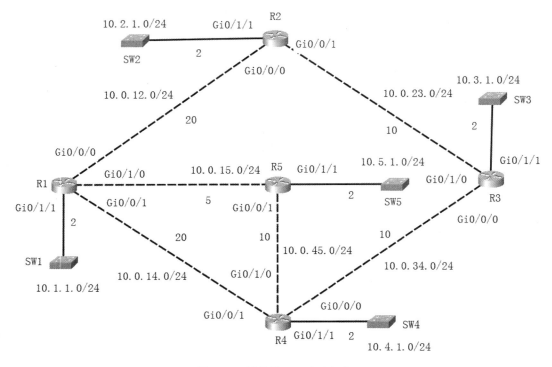

图 3-20 挑战练习一实验拓扑

任务要求：

- 请按照图 3-20 所示搭建网络拓扑；
- 请配置设备主机名和接口 IP 地址；
- 请配置单区域 OSPFv2，使全网互通；
- 请根据链路提供的数值配置链路开销；
- 在所有交换机上添加相应 PC 并配置地址参数；
- 请在 PC 上使用 **tracert** 命令跟踪与远程主机通信的路径；
- 请检验实验跟踪路径与你依据拓扑分析的结果是否一致。

3.5.2 挑战练习二

趣味屋：挑战练习二实验拓扑如图 3-21 所示，请根据所学知识完成 OSPFv2 配置，确保全网连通。

请读者独立完成该挑战练习。

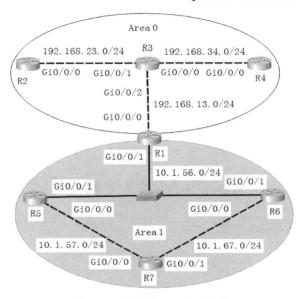

图 3-21　挑战练习二实验拓扑

任务要求：

- 请按照图 3-21 所示搭建网络拓扑；
- 请配置设备主机名和接口 IP 地址；
- 请配置多区域 OSPFv2，指定路由器 R*x* 的路由器 ID 为 X.X.X.X（*x*=1,2,3,4,5,6,7）；
- 在路由器 R1、R5 和 R6 组成的广播式网络中指定路由器 R1 为 DR，R6 没有资格参与选举；
- 更改路由器 R1 接口的 Hello 间隔和 Dead 间隔分别为 5 s 和 20 s；
- 配置路由器 R7，使其访问 Area 0 必须经过路由器 R5，路由器 R6 作为其备份下一跳；
- 将路由器 R1 通过串口接入 ISP 路由器，ISP 路由器接口地址为 221.5.27.2/30；
- 请配置路由器 R1，确保 OSPF 区域内的路由器均可访问 ISP 接口。

3.6　本章小结

本章内容到此结束。本章主要内容包括 OSPFv2 概述、基础 OSPFv2 配置、OSPFv2 检验以及 OSPFv2 的特殊选项配置。通过本章内容的学习，我们认识了 OSPFv2 以及一些重要概念，例如，数据包类型、区域类型、路由器 ID 以及路由器角色等。通过 7 个场景、5 个实验和 2 个挑战练习，我们进一步掌握了 OSPFv2 的灵活应用。在现实网络中，OSPFv2 是应用最广泛的路由协议，熟练掌握本章内容有助于我们更好地学习网络知识。

第4章 >>>

学习高级 OSPFv2

本章要点：

- 完成 OSPFv2 虚链路配置
- 完成 OSPFv2 认证配置
- 完成 OSPFv2 路由注入配置
- 完成 OSPFv2 路由汇总配置
- 完成 OSPFv2 特殊区域配置
- 挑战练习
- 本章小结

本章介绍高级 OSPFv2，是第 3 章的提高篇。其中，4.1 节介绍 OSPFv2 虚链路配置，该技术解决了骨干区域被分割以及骨干区域与非骨干区域没有直连的问题；4.2 节介绍 OSPFv2 认证配置，包括基于区域和基于接口的认证；4.3 节介绍 OSPFv2 路由注入配置，包括外部路由注入、默认路由传播以及 OSPFv2 多进程重发布；4.4 节介绍 OSPFv2 路由汇总配置，包括域间路由汇总和外部路由汇总；4.5 节介绍 OSPFv2 特殊区域配置，包括末节、完全末节、次末节以及完全次末节区域配置。本章设计了 7 个应用场景和 2 个挑战练习，以帮助读者熟练应用 OSPFv2 的高级特性，提升和优化网络性能。

4.1 完成 OSPFv2 虚链路配置

4.1.1 认识 OSPFv2 虚链路

在某些情况下，某些常规区域无法与骨干区域直连，因而也无法与其他区域建立路由，此时，可采用 OSPFv2 虚链路（Virtual Link）实现路由。OSPFv2 虚链路通过将相邻的常规区域虚拟为骨干区域，从而让那些不能与骨干区域直连的常规区域也能与其他 OSPFv2 区域建立路由。

虚链路可以为物理上未连接到主干区域的区域提供一条通向主干区域的逻辑链路。建立虚链路的要求如下：
- 虚链路必须建立在共享一个共同区域的两台路由器间；
- 这两台路由器中的其中一台必须能连接到骨干区域。

骨干区域不能有意地被分割，但是如果由于链路故障或其他原因导致区域被分割，那么可以考虑使用虚链路对骨干区域进行临时性补救。虚链路技术是一种类似于隧道技术的逻辑连接技术，但是却不是真正意义上采用多层协议封装的隧道技术。在如下情景中可以使用虚链路：
- 区域 0（骨干区域）被分割；
- 非骨干区域与区域 0 之间没有直接的物理连接。

公认的网络设计理论认为，在骨干区域或网络内使用虚链路并不是优秀的网络设计。

虚链路的特点及建议用法如下：
- 虚链路的稳定性取决于虚链路所穿越区域的稳定性；
- 虚链路只能在 ABR 上配置；
- 虚链路不能穿越末节区域；
- 虚链路可以短期性解决网络连接性故障；
- 虚链路可以提供逻辑冗余性；
- 两台路由器之间的虚链路可被看作一条未编址的点到点链路；
- 虚链路不能配置在未编址的链路上，即配置虚链路的物理链路必须具备 IP 连通性。

4.1.2 场景一：配置 OSPFv2 虚链路

经过 4.1.1 节的学习，相信大家已经对 OSPFv2 虚链路有了初步认知。接下来让我们一起通过具体实验来学习 OSPFv2 虚链路配置。本场景出现了两个 Area 0，需要使用虚链路使两个 Area 0 合并成一个。配置 OSPFv2 虚链路实验拓扑如图 4-1 所示。

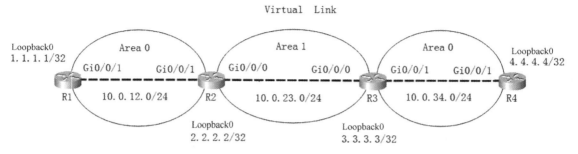

图 4-1　配置 OSPFv2 虚链路实验拓扑

任务要求：
- 按图 4-1 所示搭建网络拓扑；
- 按图 4-1 所示为 4 台路由器接口配置 IP 地址；
- 配置多区域 OSPFv2，使用虚链路技术使全网互通。

配置 OSPFv2 虚链路的语法如下（以路由器 R3 为例）：

```
R3(config)#router ospf 1
R3(config-router)#area 1 virtual-link 2.2.2.2      //在指定区域与指定路由器创建虚链路
R3#show ip ospf virtual-links                      //查看路由器虚链路配置信息
```

任务实施如下所述。

步骤一：配置路由器接口 IP 地址

（1）配置路由器 R1

```
R1(config)#interface Loopback 0
R1(config-if)#ip address 1.1.1.1 255.255.255.255
R1(config-if)#interface GigabitEthernet 0/0/1
R1(config-if)#ip address 10.0.12.1 255.255.255.0
R1(config-if)#no shutdown
```

（2）配置路由器 R2

```
R2(config)#interface Loopback 0
R2(config-if)#ip address 2.2.2.2 255.255.255.255
```

```
R2(config-if)#interface GigabitEthernet 0/0/1
R2(config-if)#ip address 10.0.12.2 255.255.255.0
R2(config-if)#no shutdown
R2(config-if)#interface GigabitEthernet 0/0/0
R2(config-if)#ip address 10.0.23.2 255.255.255.0
R2(config-if)#no shutdown
```

（3）配置路由器 R3

```
R3(config)#interface Loopback 0
R3(config-if)#ip address 3.3.3.3 255.255.255.255
R3(config-if)#interface GigabitEthernet 0/0/0
R3(config-if)#ip address 10.0.23.3 255.255.255.0
R3(config-if)#no shutdown
R3(config-if)#interface GigabitEthernet 0/0/1
R3(config-if)#ip address 10.0.34.3 255.255.255.0
R3(config-if)#no shutdown
```

（4）配置路由器 R4

```
R4(config)#interface Loopback 0
R4(config-if)#ip address 4.4.4.4 255.255.255.255
R4(config-if)#interface GigabitEthernet 0/0/1
R4(config-if)#ip address 10.0.34.4 255.255.255.0
R4(config-if)#no shutdown
```

步骤二：配置多区域 OSPFv2

（1）配置路由器 R1

```
R1(config)#router ospf 1
R1(config-router)#network 1.1.1.1 0.0.0.0 area 0
R1(config-router)#network 10.0.12.0 0.0.0.255 area 0
```

（2）配置路由器 R2

```
R2(config)#router ospf 1
R2(config-router)#network 2.2.2.2 0.0.0.0 area 0
R2(config-router)#network 10.0.12.0 0.0.0.255 area 0
R2(config-router)#network 10.0.23.0 0.0.0.255 area 1
```

（3）配置路由器 R3

```
R3(config)#router ospf 1
R3(config-router)#network 3.3.3.3 0.0.0.0 area 1
R3(config-router)#network 10.0.23.0 0.0.0.255 area 1
R3(config-router)#network 10.0.34.0 0.0.0.255 area 0
```

（4）配置路由器 R4

```
R4(config)#router ospf 1
R4(config-router)#network 4.4.4.4 0.0.0.0 area 0
R4(config-router)#network 10.0.34.0 0.0.0.255 area 0
```

步骤三：查看路由器的路由表

（1）查看路由器 R1 的路由表

```
R1>show ip route | begin Gateway
Gateway of last resort is not set
     1.0.0.0/32 is subnetted, 1 subnets
C       1.1.1.1/32 is directly connected, Loopback0
     2.0.0.0/32 is subnetted, 1 subnets
O       2.2.2.2/32 [110/2] via 10.0.12.2, 00:14:12, GigabitEthernet0/0/1
     3.0.0.0/32 is subnetted, 1 subnets
O IA    3.3.3.3/32 [110/3] via 10.0.12.2, 00:12:07, GigabitEthernet0/0/1    //区域1的主机路由
     10.0.0.0/8 is variably subnetted, 3 subnets, 2 masks
C       10.0.12.0/24 is directly connected, GigabitEthernet0/0/1
L       10.0.12.1/32 is directly connected, GigabitEthernet0/0/1
O IA    10.0.23.0/24 [110/2] via 10.0.12.2, 00:14:12, GigabitEthernet0/0/1  //区域1的网络路由
```

（2）查看路由器 R4 的路由表

```
R4>show ip route | begin Gateway
Gateway of last resort is not set
     3.0.0.0/32 is subnetted, 1 subnets
O IA    3.3.3.3/32 [110/2] via 10.0.34.3, 00:11:42, GigabitEthernet0/0/1    //区域1的主机路由
     4.0.0.0/32 is subnetted, 1 subnets
C       4.4.4.4/32 is directly connected, Loopback0
     10.0.0.0/8 is variably subnetted, 3 subnets, 2 masks
O IA    10.0.23.0/24 [110/2] via 10.0.34.3, 00:11:42, GigabitEthernet0/0/1  //区域1的网络路由
```

C	10.0.34.0/24 is directly connected, GigabitEthernet0/0/1
L	10.0.34.4/32 is directly connected, GigabitEthernet0/0/1

由以上输出可知，因为 Area 1 分割了 Area 0，导致路由表不完整，不能实现全网互通。请读者查看并分析路由器 R2 和 R3 的路由表。

步骤四：配置 OSPFv2 虚链路

（1）配置路由器 R2

```
R2(config)#router ospf 1
R2(config-router)#area 1 virtual-link 3.3.3.3
```

（2）配置路由器 R3

```
R3(config)#router ospf 1
R3(config-router)#area 1 virtual-link 2.2.2.2
00:12:10: %OSPF-5-ADJCHG: Process 1, Nbr 2.2.2.2 on OSPF_VL0 from LOADING to FULL, Loading Done
```

步骤五：查看路由器 R2 与 R3 间的虚链路

```
R2>show ip ospf virtual-links
Virtual Link OSPF_VL0 to router 3.3.3.3 is up
    Run as demand circuit
    Transit area 1, via interface GigabitEthernet0/0/0, Cost of using 1
    Transmit Delay is 1 sec, State POINT_TO_POINT,
    Timer intervals configured, Hello 10, Dead 40, Wait 40, Retransmit 5
        Hello due in 00:00:08
    Adjacency State FULL
        Index 1/2, retransmission queue length 0, number of retransmission 0
        First 0x0(0)/0x0(0) Next 0x0(0)/0x0(0)
        Last retransmission scan length is 0, maximum is 0
        Last retransmission scan time is 0 msec, maximum is 0 msec
```

步骤六：查看路由表的变化并测试连通性

以路由器 R1 为例，查看其路由表。

```
R1>show ip route | begin Gateway
Gateway of last resort is not set
        1.0.0.0/32 is subnetted, 1 subnets
C          1.1.1.1/32 is directly connected, Loopback0
```

```
                2.0.0.0/32 is subnetted, 1 subnets
O                  2.2.2.2/32 [110/2] via 10.0.12.2, 00:55:53, GigabitEthernet0/0/1
                3.0.0.0/32 is subnetted, 1 subnets
O IA            3.3.3.3/32 [110/3] via 10.0.12.2, 00:55:43, GigabitEthernet0/0/1
                4.0.0.0/32 is subnetted, 1 subnets
O               4.4.4.4/32 [110/4] via 10.0.12.2, 00:03:13, GigabitEthernet0/0/     //添加 Area 0 的域内路由
                10.0.0.0/8 is variably subnetted, 4 subnets, 2 masks
C                  10.0.12.0/24 is directly connected, GigabitEthernet0/0/1
L                  10.0.12.1/32 is directly connected, GigabitEthernet0/0/1
O IA            10.0.23.0/24 [110/2] via 10.0.12.2, 00:55:53, GigabitEthernet0/0/1
O               10.0.34.0/24 [110/3] via 10.0.12.2, 00:03:13, GigabitEthernet0/0/1  //添加 Area 0 的域内路由
```

以路由器 R1 为例，测试连通性。

```
R1>ping 4.4.4.4
Sending 5, 100-byte ICMP Echos to 4.4.4.4, timeout is 2 seconds:
!!!!!
Success rate is 100 percent (5/5), round-trip min/avg/max = 0/0/0 ms
```

通过场景一实验可以发现，当未配置 OSPFv2 虚链路时，路由器 R1 与 R4 间无法通信；当配置完虚链路后，通信正常。

接下来我们通过图 4-2 所示的自测实验拓扑，趁热打铁，检验读者对虚链路技术的应用情况。请读者独立完成以下自测实验，图中有 3 个区域，Area 2 与 Area 0 直连，Area 1 通过 Area 2 间接与 Area 0 连接。

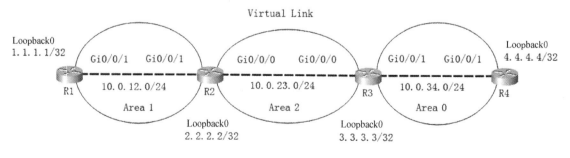

图 4-2　自测实验拓扑

任务要求：
- 按图 4-2 搭建网络拓扑；
- 按图 4-2 为路由器接口配置 IP 地址；
- 配置多区域 OSPFv2，使用虚链路使全网互通。

4.2 完成 OSPFv2 认证配置

4.2.1 认识 OSPFv2 认证

OSPFv2 的首要职责并不是对网络数据流进行转发，而是通过交互路由更新信息来构建路由表。因此，OSPFv2 所提供的认证特性对于保护路由协议的自身功能而言，已然足够。为了提供路由认证功能，OSPFv2 在其数据包中预留了所需要的安全字段。但是，这些安全字段仅用于保护 OSPFv2 的邻居关系和链路状态通告（LSA），维持网络路由信息的完整性。

OSPFv2 具有最小化的安全性，也就是说，这种安全性并不能保护穿越网络的数据流量，只能用于确保配置 OSPFv2 的路由器掌握准确有效的路由信息。OSPFv2 的安全性只能维持 OSPFv2 路由域内路由信息的完整性。

OSPFv2 认证的操作特点：
- OSPFv2 认证可在单条链路或整个 OSPFv2 区域完成；
- 位于同一条链路上的邻居路由器所使用的认证密钥必须匹配；
- OSPFv2 明文认证几乎不堪一击，因此尽可能选择使用 MD5 认证。

4.2.2 场景二：配置 OSPFv2 安全认证

经过 4.2.1 节的学习，相信大家已经了解关于 OSPFv2 认证的相关知识。接下来让我们一起通过具体实验完成 OSPFv2 认证配置。OSPFv2 认证配置实验拓扑如图 4-3 所示。

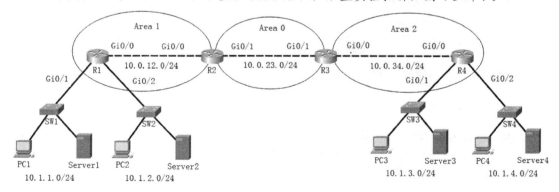

图 4-3 OSPFv2 认证配置实验拓扑

任务要求：
- 按图 4-3 所示搭建网络拓扑；
- 按图 4-3 中所示为设备配置 IP 地址；
- 完成 OSPFv2 认证配置，使全网互通。

➤ 路由器 R1 和 R2 之间采用基于区域的明文认证，密钥为 cisco；

➤ 路由器 R3 和 R4 之间采用基于接口的 MD5 认证，密钥为 cisco。

配置 OSPFv2 认证语法如下。

① 基于区域的 OSPFv2 认证（以路由器 R1 为例）。

基于区域的明文认证：

  R1(config-router)#**area 1 authentication**

  R1(config-router)#**interface GigabitEthernet 0/0**

  R1(config-if)#**ip ospf authentication-key cisco**

基于区域的 MD5 认证：

  R1(config-router)#**area 1 authentication message-digest**

  R1(config-router)#**interface GigabitEthernet 0/0**

  R1(config-if)#**ip ospf message-digest-key 1 md5 cisco**

② 基于接口的 OSPFv2 认证（以路由器 R3 为例）。

基于接口的明文认证：

  R3(config-router)#**interface GigabitEthernet 0/0**

  R3(config-if)#**ip ospf authentication**

  R3(config-if)#**ip ospf authentication-key cisco**

基于接口的 MD5 认证：

  R3(config-router)#**interface GigabitEthernet 0/0**

  R3(config-if)#**ip ospf authentication message-diges**t

  R3(config-if)#**ip ospf message-digest-key 1 md5 cisco**

任务实施如下所述。

步骤一：配置路由器接口 IP 地址

（1）配置路由器 R1

  R1(config)#**interface GigabitEthernet 0/1**

  R1(config-if)#**ip address 10.1.1.254 255.255.255.0**

  R1(config-if)#**no shutdown**

  R1(config-if)#**interface GigabitEthernet 0/2**

  R1(config-if)#**ip address 10.1.2.254 255.255.255.0**

  R1(config-if)#**no shutdown**

  R1(config-if)#**interface GigabitEthernet 0/0**

  R1(config-if)#**ip address 10.0.12.1 255.255.255.0**

  R1(config-if)#**no shutdown**

（2）配置路由器 R2

```
R2(config)#interface GigabitEthernet 0/0
R2(config-if)#ip address 10.0.12.2 255.255.255.0
R2(config-if)#no shutdown
R2(config-if)#interface GigabitEthernet 0/1
R2(config-if)#ip address 10.0.23.2 255.255.255.0
R2(config-if)#no shutdown
```

（3）配置路由器 R3

```
R3(config)#interface GigabitEthernet 0/0
R3(config-if)#ip address 10.0.34.3 255.255.255.0
R3(config-if)#no shutdown
R3(config-if)#interface GigabitEthernet 0/1
R3(config-if)#ip address 10.0.23.3 255.255.255.0
R3(config-if)#no shutdown
```

（4）配置路由器 R4

```
R4(config)#interface GigabitEthernet 0/1
R4(config-if)#ip address 10.1.3.254 255.255.255.0
R4(config-if)#no shutdown
R4(config-if)#interface GigabitEthernet 0/2
R4(config-if)#ip address 10.1.4.254 255.255.255.0
R4(config-if)#no shutdown
R4(config-if)#interface GigabitEthernet 0/0
R4(config-if)#ip address 10.0.34.4 255.255.255.0
R4(config-if)#no shutdown
```

步骤二：配置 OSPFv2 并完成认证

（1）配置路由器 R1

```
R1(config)#router ospf 1
R1(config-router)#router-id 1.1.1.1
R1(config-router)#network 10.1.1.0 0.0.0.255 area 1
R1(config-router)#network 10.1.2.0 0.0.0.255 area 1
R1(config-router)#network 10.0.12.0 0.0.0.255 area 1
R1(config-router)#area 1 authentication                //基于区域的明文认证
R1(config-router)#interface GigabitEthernet 0/0
```

R1(config-if)#**ip ospf authentication-key cisco**

（2）配置路由器 R2

```
R2(config)#router ospf 1
R2(config-router)#router-id 2.2.2.2
R2(config-router)#network 10.0.12.0 0.0.0.255 area 1
R2(config-router)#network 10.0.23.0 0.0.0.255 area 0
R2(config-router)#area 1 authentication              //基于区域的明文认证
R2(config-router)#interface GigabitEthernet 0/0
R2(config-if)#ip ospf authentication-key cisco
```

（3）配置路由器 R3

```
R3(config)#router ospf 1
R3(config-router)#router-id 3.3.3.3
R3(config-router)#network 10.0.23.0 0.0.0.255 area 0
R3(config-router)#network 10.0.34.0 0.0.0.255 area 2
R3(config-router)#interface GigabitEthernet 0/0
R3(config-if)#ip ospf authentication message-digest   //基于接口的 MD5 认证
R3(config-if)#ip ospf message-digest-key 1 md5 cisco
```

（4）配置路由器 R4

```
R4(config)#router ospf 1
R4(config-router)#router-id 4.4.4.4
R4(config-router)#network 10.0.34.0 0.0.0.255 area 2
R4(config-router)#network 10.1.3.0 0.0.0.255 area 2
R4(config-router)#network 10.1.4.0 0.0.0.255 area 2
R4(config-router)#interface Gigabitethernet 0/0
R4(config-if)#ip ospf authentication message-digest   //基于接口的 MD5 认证
R4(config-if)#ip ospf message-digest-key 1 md5 cisco
```

至此，我们通过场景二的实验完成了 OSPFv2 认证配置，包括基于区域的明文认证和基于接口的 MD5 认证，希望能帮助读者进一步理解 OSPFv2 认证对网络环境所起的安全作用。

4.3 完成 OSPFv2 路由注入配置

4.3.1 认识外部路由注入

外部路由是指信源和信宿位于 OSPFv2 的不同 AS 内的路由。获得路由的最为常用的方法是将路由重分布到 OSPFv2 区域内。为使所有的 OSPFv2 路由器获知外部路由信息，自治系统边界路由器（ASBR）负责泛洪外部路由信息到 AS 内，但默认不会对这些路由进行路由汇总。默认情况下，AS 内的所有路由器都可以接收到外部路由信息。特别值得一提的是，OSPFv2 拥有两种外部路由类型：E1 路由、E2 路由。下面我们将来简单介绍这两种外部路由，E1 路由和 E2 路由的特点如表 4-1 所示。

表 4-1　E1 路由和 E2 路由的特点

外部路由类型	路由表中标识符	OSPFv2 外部路由度量的计算方法	度量程度
E1 路由	O E1	内部度量+外部 OSPFv2 度量（被重分布到 OSPFv2 区域的路由的初始度量）	精确度量（关注区域内路由器的位置）
E2 路由	O E2	仅用外部 OSPFv2 度量，不计算内部度量，是 OSPFv2 默认外部路由类型	粗糙度量（不关注区域内路由器的位置）

4.3.2　场景三：配置外部路由注入及默认路由传播

经过前面理论知识的学习，相信大家已经对外部路由注入有了初步认知。接下来，让我们一起通过具体实验学习如何完成外部路由注入及默认路由传播配置。外部路由注入配置实验拓扑如图 4-4 所示。

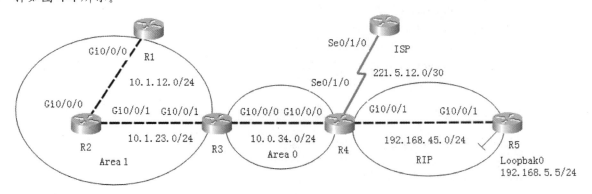

图 4-4　外部路由注入配置实验拓扑

任务要求：

- 按图 4-4 所示搭建网络拓扑；
- 按图 4-4 所示为路由器接口配置 IP 地址；
- 按图 4-4 所示配置路由协议 OSPFv2 和 RIPv2；
- 在路由器 R4 上配置静态默认路由并传播至 OSPFv2 区域和 RIPv2 区域；
- 在路由器 R4 上将 RIPv2 外部路由注入 OSPFv2 区域；
- 测试网络连通性。

外部路由注入语法如下（以路由器 R4 为例）：

R4(config-router)#**router ospf 1**
R4(config-router)#**redistribute rip** [subnets]   //将 RIP 路由注入 OSPFv2 区域

传播默认路由语法如下（以路由器 R4 为例）：

R4(config-router)#**router ospf 1**
R4(config-router)#**default-information originate**

任务实施如下所述。

步骤一：配置路由器接口 IP 地址

（1）配置路由器 R1

R1(config)#**interface GigabitEthernet 0/0/0**
R1(config-if)#**ip address 10.1.12.1 255.255.255.0**
R1(config-if)#**no shutdown**

（2）配置路由器 R2

R2(config)#**interface GigabitEthernet 0/0/0**
R2(config-if)#**ip address 10.1.12.2 255.255.255.0**
R2(config-if)#**no shutdown**
R2(config-if)#**interface GigabitEthernet 0/0/1**
R2(config-if)#**ip address 10.1.23.2 255.255.255.0**
R2(config-if)#**no shutdown**

（3）配置路由器 R3

R3(config)#**interface GigabitEthernet 0/0/0**
R3(config-if)#**ip address 10.0.34.3 255.255.255.0**
R3(config-if)#**no shutdown**
R3(config-if)#**interface GigabitEthernet 0/0/1**
R3(config-if)#**ip address 10.1.23.3 255.255.255.0**
R3(config-if)#**no shutdown**

（4）配置路由器 R4

R4(config)#**interface GigabitEthernet 0/0/0**
R4(config-if)#**ip address 10.0.34.4 255.255.255.0**
R4(config-if)#**no shutdown**
R4(config-if)#**interface GigabitEthernet 0/0/1**
R4(config-if)#**ip address 192.168.45.4 255.255.255.0**

```
R4(config-if)#no shutdown
R4(config-if)#interface Serial0/1/0
R4(config-if)#ip address 221.5.12.1 255.255.255.0
R4(config-if)#no shutdown
```

（5）配置路由器 R5

```
R5(config)#interface GigabitEthernet 0/0/1
R5(config-if)#ip address 192.168.45.5 255.255.255.0
R5(config-if)#no shutdown
R5(config-if)#interface Loopback 0
R5(config-if)#ip   address 192.168.5.5 255.255.255.0
```

（6）配置路由器 ISP

```
ISP(config)#interface Serial0/1/0
ISP(config-if)#ip address 221.5.12.2 255.255.255.0
ISP(config-if)#no shutdown
```

步骤二：配置 OSPFv2 与 RIPv2

（1）配置路由器 R1

```
R1(config)#router ospf 1
R1(config-router)#router-id 1.1.1.1
R1(config-router)#network 10.1.12.0 0.0.0.255 area 1
```

（2）配置路由器 R2

```
R2(config)#router ospf 1
R2(config-router)#router-id 2.2.2.2
R2(config-router)#network 10.1.12.0 0.0.0.255 area 1
R2(config-router)#network 10.1.23.0 0.0.0.255 area 1
```

（3）配置路由器 R3

```
R3(config)#router ospf 1
R3(config-router)#router-id 3.3.3.3
R3(config-router)#network 10.1.23.0 0.0.0.255 area 1
R3(config-router)#network 10.0.34.0 0.0.0.255 area 0
```

（4）配置路由器 R4

```
R4(config)#router ospf 1
```

R4(config-router)#**router-id 4.4.4.4**
R4(config-router)#**network 10.0.34.0 0.0.0.255 area 0**
R4(config-router)#**router rip**
R4(config-router)#**version 2**
R4(config-router)#**network 192.168.45.0**

（5）配置路由器 R5

R5(config)#**router rip**
R5(config-router)#**version 2**
R5(config-router)#**network 192.168.45.0**
R5(config-router)#**network 192.168.5.0**

步骤三：查看路由器的路由表

R1>**show ip route | begin Gateway**
Gateway of last resort is not set

    10.0.0.0/8 is variably subnetted, 4 subnets, 2 masks
O IA   **10.0.34.0/24 [110/3] via 10.1.12.2, 00:02:57, GigabitEthernet0/0/0**
C   10.1.12.0/24 is directly connected, GigabitEthernet0/0/0
L   10.1.12.1/32 is directly connected, GigabitEthernet0/0/0
O   **10.1.23.0/24 [110/2] via 10.1.12.2, 00:02:57, GigabitEthernet0/0/0**

由以上输出可知，路由器 R1 添加了 10.1.23.0/24 和 10.0.34.0/24 的 OSPFv2 区域的路由。

R5>**show ip route | begin Gateway**
Gateway of last resort is not set

    192.168.5.0/24 is variably subnetted, 2 subnets, 2 masks
C   **192.168.5.0/24 is directly connected, Loopback0**
L   192.168.5.5/32 is directly connected, Loopback0
    192.168.45.0/24 is variably subnetted, 2 subnets, 2 masks
C   **192.168.45.0/24 is directly connected, GigabitEthernet0/0/1**
L   192.168.45.5/32 is directly connected, GigabitEthernet0/0/1

由以上输出可知，路由器 R5 仅添加了直连路由。我们进一步查看路由器 R2、R3 和 R4 的路由表，发现 R4 添加了所有网段路由，路由器 R2、R3 与 R1 一样添加了 OSPFv2 区域的所有路由。

步骤四：在 ASBR（路由器 R4）上完成默认路由传播和外部路由注入配置

（1）配置静态默认路由

R4(config)#**ip route 0.0.0.0 0.0.0.0 Serial 0/1/0**

（2）配置传播默认路由

```
R4(config)#router ospf 1
R4(config-router)#default-information originate    //传播默认路由信息至 OSPFv2 区域
R4(config-router)#router rip
R4(config-router)#default-information originate    //传播默认路由信息至 RIP 区域
```

（3）配置外部路由注入

```
R4(config)#router ospf 1
R4(config-router)#redistribute rip
```

步骤五：查看路由器路由表的变化

```
R1>show ip route | begin Gateway
Gateway of last resort is 10.1.12.2 to network 0.0.0.0
        10.0.0.0/8 is variably subnetted, 4 subnets, 2 masks
O IA    10.0.34.0/24 [110/3] via 10.1.12.2, 00:05:15, GigabitEthernet0/0/0
C       10.1.12.0/24 is directly connected, GigabitEthernet0/0/0
L       10.1.12.1/32 is directly connected, GigabitEthernet0/0/0
O       10.1.23.0/24 [110/2] via 10.1.12.2, 00:05:15, GigabitEthernet0/0/0
O E2 192.168.5.0/24 [110/20] via 10.1.12.2, 00:00:24, GigabitEthernet0/0/0
O E2 192.168.45.0/24 [110/20] via 10.1.12.2, 00:00:24, GigabitEthernet0/0/0
O*E2 0.0.0.0/0 [110/1] via 10.1.12.2, 00:05:15, GigabitEthernet0/0/0
```

由以上输出可知，路由器 R1 添加了 RIP 区域的外部路由和从外部学习来的默认路由。

```
R4>show ip route | begin Gateway
Gateway of last resort is 0.0.0.0 to network 0.0.0.0
        10.0.0.0/8 is variably subnetted, 4 subnets, 2 masks
C       10.0.34.0/24 is directly connected, GigabitEthernet0/0/0
L       10.0.34.4/32 is directly connected, GigabitEthernet0/0/0
O IA    10.1.12.0/24 [110/3] via 10.0.34.3, 00:05:57, GigabitEthernet0/0/0
O IA    10.1.23.0/24 [110/2] via 10.0.34.3, 00:05:57, GigabitEthernet0/0/0
R       192.168.5.0/24 [120/1] via 192.168.45.5, 00:00:24, GigabitEthernet0/0/1
        192.168.45.0/24 is variably subnetted, 2 subnets, 2 masks
C       192.168.45.0/24 is directly connected, GigabitEthernet0/0/1
L       192.168.45.4/32 is directly connected, GigabitEthernet0/0/1
        221.5.12.0/24 is variably subnetted, 2 subnets, 2 masks
C       221.5.12.0/24 is directly connected, Serial0/1/0
L       221.5.12.1/32 is directly connected, Serial0/1/0
```

```
S*    0.0.0.0/0 is directly connected, Serial0/1/0
```

由以上输出可知，路由器 R4 添加了通往 ISP 的静态默认路由，其路由表完整。

```
R5>show ip route | begin Gateway
Gateway of last resort is 192.168.45.4 to network 0.0.0.0

      192.168.5.0/24 is variably subnetted, 2 subnets, 2 masks
C        192.168.5.0/24 is directly connected, Loopback0
L        192.168.5.5/32 is directly connected, Loopback0
      192.168.45.0/24 is variably subnetted, 2 subnets, 2 masks
C        192.168.45.0/24 is directly connected, GigabitEthernet0/0/1
L        192.168.45.5/32 is directly connected, GigabitEthernet0/0/1
R*    0.0.0.0/0 [120/1] via 192.168.45.4, 00:00:12, GigabitEthernet0/0/1
```

由以上输出可知，路由器 R5 添加了从由路由器 R4 学习来的默认路由。

步骤六：测试连通性

请依次进行连通性测试。以下展示了路由器 R1 到 R5 环回接口的连通性测试结果——通信成功。

```
R1>ping 192.168.5.5
Sending 5, 100-byte ICMP Echos to 192.168.5.5, timeout is 2 seconds:
!!!!!
Success rate is 100 percent (5/5), round-trip min/avg/max = 0/0/0 ms
```

以下展示了路由器 R1 到 R5 环回接口的数据包跟踪路径。

```
R1>traceroute 192.168.5.5
Tracing the route to 192.168.5.5
  1   10.1.12.2      0 msec   0 msec   0 msec
  2   10.1.23.3      0 msec   0 msec   0 msec
  3   10.0.34.4      0 msec   0 msec   0 msec
  4   192.168.45.5   0 msec   0 msec   0 msec
```

以上的数据包跟踪路径为经 R2、R3 和 R4 到达目标路由器 R5。请思考：在本场景中，我们将 RIPv2 区域的路由注入 OSPFv2 区域，没有实现 OSPFv2 区域路由注入 RIPv2 区域，为什么可以实现 RIPv2 区域和 OSPFv2 区域间的互通？如果不将 RIPv2 区域路由注入 OSPFv2 区域，能否实现互通？请说明理由。

4.3.3　场景四：配置多进程 OSPFv2

经过 4.3.1 节和 4.3.2 节内容的学习，我们了解了采用 OSPFv2 如何处理不同 AS 之间的问题，即通过路由注入或称为路由重发布实现不同 AS 之间的通信。接下来,让我们一起通过具体实验来学习如何进行 OSPFv2 多进程发布配置。OSPFv2 多进程发布配置实验拓扑如图 4-5 所示。

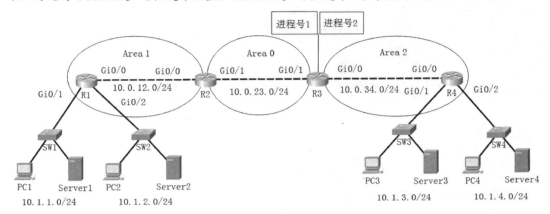

图 4-5　OSPFv2 多进程发布配置实验拓扑

任务要求:
- 按图 4-5 所示搭建拓扑;
- 按图 4-5 所示为设备配置 IP 地址;
- 按图 4-5 所示配置 OSPFv2 多进程;
- 配置 OSPFv2 多进程重发布使得全网互通。

OSPFv2 多进程重发布语法如下(以路由器 R3 为例):

```
R3(config-router)#router ospf 1
R3(config-router)#redistribute ospf 2 subnets    //将 OSPFv2 进程 2 中的路由注入进程 1
```

任务实施如下所述。

步骤一:配置路由器接口 IP 地址

(1)配置路由器 R1

```
R1(config)#interface GigabitEthernet 0/0
R1(config-if)#ip address 10.0.12.1 255.255.255.0
R1(config-if)#no shutdown
R1(config-if)#interface GigabitEthernet 0/1
R1(config-if)#ip address 10.1.1.254 255.255.255.0
R1(config-if)#no shutdown
R1(config-if)#interface GigabitEthernet 0/2
R1(config-if)#ip address 10.1.2.254 255.255.255.0
R1(config-if)#no shutdown
```

（2）配置路由器 R2

```
R2(config)#interface GigabitEthernet 0/0
R2(config-if)#ip address 10.0.12.2 255.255.255.0
R2(config-if)#no shutdown
R2(config-if)#interface GigabitEthernet 0/1
R2(config-if)#ip address 10.0.23.2 255.255.255.0
R2(config-if)#no shutdown
```

（3）配置路由器 R3

```
3(config)#interface GigabitEthernet 0/0
R3(config-if)#ip address 10.0.34.3 255.255.255.0
R3(config-if)#no shutdown
R3(config-if)#interface GigabitEthernet 0/1
R3(config-if)#ip address 10.0.23.3 255.255.255.0
R3(config-if)#no shutdown
```

（4）配置路由器 R4

```
R4(config)#interface GigabitEthernet 0/0
R4(config-if)#ip address 10.0.34.4 255.255.255.0
R4(config-if)#no shutdown
R4(config-if)#interface GigabitEthernet 0/1
R4(config-if)#ip address 10.1.3.254 255.255.255.0
R4(config-if)#no shutdown
R4(config-if)#interface GigabitEthernet 0/2
R4(config-if)#ip address 10.1.4.254 255.255.255.0
R4(config-if)#no shutdown
```

步骤二：配置多进程 OSPFv2

如图 4-5 所示配置 OSPFv2，注意路由器 R3 的两个 Area 所用的进程号不同。

（1）配置路由器 R1

```
R1(config)#router ospf 1
R1(config-router)#router-id 1.1.1.1
R1(config-router)#network 10.1.1.0 0.0.0.255 area 1
R1(config-router)#network 10.1.2.0 0.0.0.255 area 1
R1(config-router)#network 10.0.12.0 0.0.0.255 area 1
```

（2）配置路由器 R2

> R2(config)#**router ospf 1**
> R2(config-router)#**router-id 2.2.2.2**
> R2(config-router)#**network 10.0.12.0 0.0.0.255 area 1**
> R2(config-router)#**network 10.0.23.0 0.0.0.255 area 0**

（3）配置路由器 R3

> R3(config)#**router ospf 1**
> R3(config-router)#**router-id 3.3.3.3**
> R3(config-router)#**network 10.0.23.0 0.0.0.255 area 0**
> R3(config-router)#**router ospf 2**
> R3(config-router)#**router-id 3.3.2.3**
> R3(config-router)#**network 10.0.34.0 0.0.0.255 area 2**

（4）配置路由器 R4

> R4(config)#**router ospf 2**
> R4(config-router)#**router-id 4.4.4.4**
> R4(config-router)#**network 10.0.34.0 0.0.0.255 area 2**
> R4(config-router)#**network 10.1.3.0 0.0.0.255 area 2**
> R4(config-router)#**network 10.1.4.0 0.0.0.255 area 2**

步骤三：查看路由器的路由表

> R1>**show ip route | begin 10.0.0.0**
>   10.0.0.0/8 is variably subnetted, 7 subnets, 2 masks
> C 10.0.12.0/24 is directly connected, GigabitEthernet0/0
> L 10.0.12.1/32 is directly connected, GigabitEthernet0/0
> O IA 10.0.23.0/24 [110/2] via 10.0.12.2, 00:21:46, GigabitEthernet0/0
> C 10.1.1.0/24 is directly connected, GigabitEthernet0/1
> L 10.1.1.254/32 is directly connected, GigabitEthernet0/1
> C 10.1.2.0/24 is directly connected, GigabitEthernet0/2
> L 10.1.2.254/32 is directly connected, GigabitEthernet0/2

由以上输出可知，路由器 R1 没有到达 10.0.34.0/24、10.1.3.0/24 和 10.1.4.0/24 的路由。

> R4>**show ip route | begin 10.0.0.0**
>   10.0.0.0/8 is variably subnetted, 6 subnets, 2 masks
> C 10.0.34.0/24 is directly connected, GigabitEthernet0/0
> L 10.0.34.4/32 is directly connected, GigabitEthernet0/0

```
C        10.1.3.0/24 is directly connected, GigabitEthernet0/1
L        10.1.3.254/32 is directly connected, GigabitEthernet0/1
C        10.1.4.0/24 is directly connected, GigabitEthernet0/2
L        10.1.4.254/32 is directly connected, GigabitEthernet0/2
```

由以上输出可知，路由器 R4 没有到 10.0.12.0/24、10.0.23.0/24、10.1.1.0/24 和 10.1.2.0/24 的路由。我们进一步查看路由器 R2 和 R3 的路由表，发现 R3 的路由表中添加了所有网段的路由，R2 的路由表中仅添加了 2 条域内路由 10.1.1.0/24 和 10.1.2.0/24。

步骤四：在路由器 R3 上配置 OSPFv2 多进程重发布

```
R3(config)#router ospf 1
R3(config-router)#redistribute ospf 2 subnets    //将 OSPFv2 进程 2 中的路由发布到进程 1 中
R3(config-router)#router ospf 2
R3(config-router)#redistribute ospf 1 subnets    //将 OSPFv2 进程 1 中的路由发布到进程 2 中
```

步骤五：查看路由器的路由表变化

完成 OSPFv2 多进程重发布后，接下来我们查看路由器路由表的变化。

```
R1>show ip route | begin 10.0.0.0
         10.0.0.0/8 is variably subnetted, 10 subnets, 2 masks
C        10.0.12.0/24 is directly connected, GigabitEthernet0/0
L        10.0.12.1/32 is directly connected, GigabitEthernet0/0
O IA     10.0.23.0/24 [110/2] via 10.0.12.2, 01:11:28, GigabitEthernet0/0
O E2     10.0.34.0/24 [110/20] via 10.0.12.2, 00:00:21, GigabitEthernet0/0
C        10.1.1.0/24 is directly connected, GigabitEthernet0/1
L        10.1.1.254/32 is directly connected, GigabitEthernet0/1
C        10.1.2.0/24 is directly connected, GigabitEthernet0/2
L        10.1.2.254/32 is directly connected, GigabitEthernet0/2
O E2     10.1.3.0/24 [110/20] via 10.0.12.2, 00:00:21, GigabitEthernet0/0
O E2     10.1.4.0/24 [110/20] via 10.0.12.2, 00:00:21, GigabitEthernet0/0
```

由以上输出可知，路由器 R1 已经添加到 10.0.34.0/24、10.1.3.0/24 和 10.1.4.0/24 的外部路由。

```
R4>show ip route | begin 10.0.0.0
         10.0.0.0/8 is variably subnetted, 10 subnets, 2 masks
O E2     10.0.12.0/24 [110/20] via 10.0.34.3, 00:01:13, GigabitEthernet0/0
O E2     10.0.23.0/24 [110/20] via 10.0.34.3, 00:01:13, GigabitEthernet0/0
C        10.0.34.0/24 is directly connected, GigabitEthernet0/0
L        10.0.34.4/32 is directly connected, GigabitEthernet0/0
O E2     10.1.1.0/24 [110/20] via 10.0.34.3, 00:01:13, GigabitEthernet0/0
```

```
O E2     10.1.2.0/24 [110/20] via 10.0.34.3, 00:01:13, GigabitEthernet0/0
C        10.1.3.0/24 is directly connected, GigabitEthernet0/1
L        10.1.3.254/32 is directly connected, GigabitEthernet0/1
C        10.1.4.0/24 is directly connected, GigabitEthernet0/2
L        10.1.4.254/32 is directly connected, GigabitEthernet0/2
```

由以上输出可知，路由器 R4 添加了 4 条通往区域 2 的外部路由。在本场景中，我们验证了 OSPFv2 将不同进程的 OSPFv2 路由当成外部路由处理。

步骤六：测试网络的连通性

此时，全网实现可以实现互通，请依次进行连通性测试。

```
C:\>tracert 10.1.4.1
Tracing route to 10.1.4.1 over a maximum of 30 hops:
  1   0 ms   0 ms   0 ms   10.1.1.254
  2   0 ms   0 ms   0 ms   10.0.12.2
  3   0 ms   0 ms   0 ms   10.0.23.3
  4   0 ms   0 ms   0 ms   10.0.34.4
  5   0 ms   0 ms   0 ms   10.1.4.1
Trace complete.
```

以上显示了从主机 PC1 发给 PC4 的数据包的跟踪路径，该数据包经历了 4 跳到达目标。

4.4 完成 OSPFv2 路由汇总配置

4.4.1 认识 OSPFv2 路由汇总

路由汇总是把多条路由聚合成单条路由。在 OSPFv2 中，路由汇总只能在 ABR 或 ASBR 上执行。尽管可以在区域之间配置任意方向的路由汇总，但是推荐汇总那些通告到骨干区域的域间路由。这样，OSPFv2 骨干区域接收到的域间路由将全部统一为聚合路由，这些聚合路由再被注入其他区域。如果网络使用 OSPFv2 作为路由协议，那么可以实现外部路由汇总和域间路由汇总两种类型的路由汇总。

（1）外部路由汇总

为了简化重分布或控制重分布路由，可以汇总通告 OSPFv2 重分布所导入的外部路由。

（2）域间路由汇总

OSPFv2 不同区域之间的路由选择被称为域间路由，换言之，通告其他 OSPFv2 区域内的路由被称为域间路由。汇总 OSPFv2 域间路由，可优化路由选择，减少被汇总区域内的路由抖动对其他区域造成的影响。

（3）路由汇总特点

- 路由汇总是将多条明细路由聚合成单条路由后通告；
- 路由汇总会减少路由表中路由条数，提高查询效率；
- 路由汇总可以减少 LSA 泛洪，节省 CPU 和内存资源；
- OSPFv2 路由汇总只能在 ABR 或 ASBR 路由器上执行。

无论是进行哪种类型的路由汇总，OSPFv2 都会产生一条汇总 LSA（3 类 LSA 或 5 类 LSA）并将其泛洪给区域 0（骨干区域），然后再由骨干区域将汇总 LSA 转发到其他 OSPFv2 区域。需要注意的是，ASBR 只能汇总自己导入的外部路由，因为其他路由器没有资格更改不是自己创建的 LSA。

（4）两种类型的路由汇总的区别

两种类型的路由汇总的区别如表 4-2 所示。

表 4-2　两种类型的路由汇总的区别

汇总类型	产生 LSA	LSA 源自	路由标识
域间路由汇总	3 类 LSA	ABR	O IA
外部路由汇总	5 类 LSA	ASBR	O E

4.4.2　场景五：配置 OSPFv2 域间路由汇总

经过对 OSPFv2 路由汇总理论知识的学习，相信大家已经对 OSPFv2 路由汇总有了初步理解，接下来，让我们一起通过具体实验完成 OSPFv2 域间路由汇总配置。OSPFv2 域间路由汇总配置实验拓扑如图 4-6 所示。

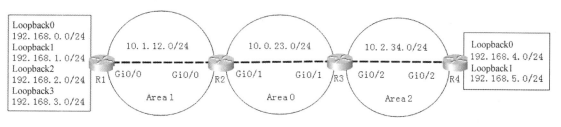

图 4-6　OSPFv2 域间路由汇总配置实验拓扑

任务要求：

- 按图 4-6 所示搭建网络拓扑；
- 按图 4-6 所示为路由器接口配置 IP 地址；

- 按图 4-6 所示配置动态路由协议 OSPFv2，使全网互通；
- 配置 OSPFv2 域间路由汇总，优化路由表，提升网络性能。

OSPFv2 域间路由汇总的命令语法如下（以路由器 R3 为例）：

```
R3(config)#router ospf 1
R3(config-router)#area 2 range 192.168.4.0 255.255.254.0    //将区域 2 的网络地址进行汇总
```

任务实施如下所述。

步骤一：配置路由器接口 IP 地址

（1）配置路由器 R1

```
R1(config)#interface Loopback0
R1(config-if)#ip address 192.168.0.1 255.255.255.0
R1(config-if)#interface Loopback1
R1(config-if)#ip address 192.168.1.1 255.255.255.0
R1(config-if)#interface Loopback2
R1(config-if)#ip address 192.168.2.1 255.255.255.0
R1(config-if)#interface Loopback3
R1(config-if)#ip address 192.168.3.1 255.255.255.0
R1(config-if)#interface GigabitEthernet 0/0
R1(config-if)#ip address 10.1.12.1 255.255.255.0
R1(config-if)#no shutdown
```

（2）配置路由器 R2

```
R2(config)#interface GigabitEthernet 0/0
R2(config-if)#ip address 10.1.12.2 255.255.255.0
R2(config-if)#no shutdown
R2(config-if)#interface GigabitEthernet 0/1
R2(config-if)#ip address 10.0.23.2 255.255.255.0
R2(config-if)#no shutdown
```

（3）配置路由器 R3

```
R3(config)#interface GigabitEthernet 0/1
R3(config-if)#ip address 10.0.23.3 255.255.255.0
R3(config-if)#no shutdown
R3(config-if)#interface GigabitEthernet 0/2
R3(config-if)#ip address 10.2.34.3 255.255.255.0
R3(config-if)#no shutdown
```

（4）配置路由器 R4

```
R4(config)#interface Loopback0
R4(config-if)#ip address 192.168.4.1 255.255.255.0
R4(config-if)#interface Loopback1
R4(config-if)#ip address 192.168.5.1 255.255.255.0
R4(config-if)#interface GigabitEthernet 0/2
R4(config-if)#ip address 10.2.34.4 255.255.255.0
R4(config-if)#no shutdown
```

步骤二：配置多区域 OSPFv2

如图 4-6 所示配置路由协议，实现网间互通。

（1）配置路由器 R1

```
R1(config)#router ospf 1
R1(config-router)#router-id 1.1.1.1
R1(config-router)#network 192.168.0.0 0.0.0.255 area 1
R1(config-router)#network 192.168.1.0 0.0.0.255 area 1
R1(config-router)#network 192.168.2.0 0.0.0.255 area 1
R1(config-router)#network 192.168.3.0 0.0.0.255 area 1
R1(config-router)#network 10.1.12.0 0.0.0.255 area 1
```

（2）配置路由器 R2

```
R2(config)#router ospf 1
R2(config-router)#router-id 2.2.2.2
R2(config-router)#network 10.1.12.0 0.0.0.255 area 1
R2(config-router)#network 10.0.23.0 0.0.0.255 area 0
```

（3）配置路由器 R3

```
R3(config)#router ospf 1
R3(config-router)#router-id 3.3.3.3
R3(config-router)#network 10.0.23.0 0.0.0.255 area 0
R3(config-router)#network 10.2.34.0 0.0.0.255 area 2
```

（4）配置路由器 R4

```
R4(config)#router ospf 1
R4(config-router)#router-id 4.4.4.4
```

```
R4(config-router)#network 10.2.34.0 0.0.0.255 area 2
R4(config-router)#network 192.168.4.0 0.0.0.255 area 2
R4(config-router)#network 192.168.5.0 0.0.0.255 area 2
```

步骤三：查看路由器的路由表

```
R1>show ip route | include 110
O IA     10.0.23.0/24 [110/2] via 10.1.12.2, 00:11:58, GigabitEthernet0/0
O IA     10.2.34.0/24 [110/3] via 10.1.12.2, 00:11:08, GigabitEthernet0/0
O IA     192.168.4.1/32 [110/4] via 10.1.12.2, 00:10:36, GigabitEthernet0/0
O IA     192.168.5.1/32 [110/4] via 10.1.12.2, 00:10:36, GigabitEthernet0/0
```

由以上路由器 R1 的路由表输出可知，OSPFv2 将路由器 R4 的 Loopback 接口当成主机路由处理，而不是一个网络地址。

```
R4>show ip route | include 110
O IA     10.0.23.0/24 [110/2] via 10.2.34.3, 00:13:39, GigabitEthernet0/2
O IA     10.1.12.0/24 [110/3] via 10.2.34.3, 00:13:39, GigabitEthernet0/2
O IA     192.168.0.1/32 [110/4] via 10.2.34.3, 00:13:39, GigabitEthernet0/2
O IA     192.168.1.1/32 [110/4] via 10.2.34.3, 00:13:39, GigabitEthernet0/2
O IA     192.168.2.1/32 [110/4] via 10.2.34.3, 00:13:39, GigabitEthernet0/2
O IA     192.168.3.1/32 [110/4] via 10.2.34.3, 00:13:39, GigabitEthernet0/2
```

同理，由路由器 R4 路由表可知，OSPFv2 也将路由器 R1 的 Loopback 接口当成/32 的主机路由处理。

步骤四：修改环回接口网络类型

（1）修改路由器 R1 环回接口类型

```
R1(config)#interface Loopback 0
R1(config-if)#ip ospf network point-to-point
R1(config-if)#interface Loopback1
R1(config-if)#ip ospf network point-to-point
R1(config-if)#interface Loopback2
R1(config-if)#ip ospf network point-to-point
R1(config-if)#interface Loopback3
R1(config-if)#ip ospf network point-to-point
```

（2）修改路由器 R4 环回接口类型

```
R4(config)#interface Loopback 0
```

R4(config-if)#**ip ospf network point-to-point**
R4(config-if)#**interface Loopback 1**
R4(config-if)#**ip ospf network point-to-point**

（3）查看路由器路由表的变化

```
R1>show ip route | include 110
O IA    10.0.23.0/24 [110/2] via 10.1.12.2, 00:01:36, GigabitEthernet0/0
O IA    10.2.34.0/24 [110/3] via 10.1.12.2, 00:01:36, GigabitEthernet0/0
O IA 192.168.4.0/24 [110/4] via 10.1.12.2, 00:01:26, GigabitEthernet0/0
O IA 192.168.5.0/24 [110/4] via 10.1.12.2, 00:01:26, GigabitEthernet0/0
```

由以上输出可知，在路由器 R1 的路由表中，原来/32 的主机路由，变成了/24 的网络路由。

```
R2>show ip route | include 110
O IA    10.2.34.0/24 [110/2] via 10.0.23.3, 00:02:47, GigabitEthernet0/1
O       192.168.0.0/24 [110/2] via 10.1.12.1, 00:02:52, GigabitEthernet0/0
O       192.168.1.0/24 [110/2] via 10.1.12.1, 00:02:52, GigabitEthernet0/0
O       192.168.2.0/24 [110/2] via 10.1.12.1, 00:02:52, GigabitEthernet0/0
O       192.168.3.0/24 [110/2] via 10.1.12.1, 00:02:52, GigabitEthernet0/0
O IA 192.168.4.0/24 [110/3] via 10.0.23.3, 00:02:37, GigabitEthernet0/1
O IA 192.168.5.0/24 [110/3] via 10.0.23.3, 00:02:37, GigabitEthernet0/1
```

由以上输出可知，在路由器 R2 的路由表中添加了 R1 和 R4 的 Loopback 接口的网络路由。

```
R3>show ip route | include 110
O IA    10 .1.12.0/24 [110/2] via 10.0.23.2, 00:03:40, GigabitEthernet0/1
O IA 192.168.0.0/24 [110/3] via 10.0.23.2, 00:03:40, GigabitEthernet0/1
O IA 192.168.1.0/24 [110/3] via 10.0.23.2, 00:03:40, GigabitEthernet0/1
O IA 192.168.2.0/24 [110/3] via 10.0.23.2, 00:03:40, GigabitEthernet0/1
O IA 192.168.3.0/24 [110/3] via 10.0.23.2, 00:03:40, GigabitEthernet0/1
O       192.168.4.0/24 [110/2] via 10.2.34.4, 00:03:45, GigabitEthernet0/2
O       192.168.5.0/24 [110/2] via 10.2.34.4, 00:03:45, GigabitEthernet0/2
```

由以上输出可知，在路由器 R3 的路由表中添加了 R1 和 R4 的 Loopback 接口的网络路由。

```
R4>show ip route | include 110
O IA    10.0.23.0/24 [110/2] via 10.2.34.3, 00:04:10, GigabitEthernet0/2
O IA    10.1.12.0/24 [110/3] via 10.2.34.3, 00:04:00, GigabitEthernet0/2
O IA 192.168.0.0/24 [110/4] via 10.2.34.3, 00:04:00, GigabitEthernet0/2
O IA 192.168.1.0/24 [110/4] via 10.2.34.3, 00:04:00, GigabitEthernet0/2
O IA 192.168.2.0/24 [110/4] via 10.2.34.3, 00:04:00, GigabitEthernet0/2
O IA 192.168.3.0/24 [110/4] via 10.2.34.3, 00:04:00, GigabitEthernet0/2
```

由以上输出可知，在路由器 R4 的路由表中添加了 R1 的 Loopback 接口的网络路由。

步骤五：在 ABR 上配置域间路由汇总

（1）在路由器 R2 上配置域间路由汇总

```
R2(config)#router ospf 1
R2(config-router)#area 1 range 192.168.0.0 255.255.252.0
```

（2）在路由器 R3 上配置域间路由汇总

```
R3(config)#router ospf 1
R3(config-router)#area 2 range 192.168.4.0 255.255.254.0
```

（3）查看路由器路由表的变化

```
R1>show ip route | include 110
O IA    10.0.23.0/24 [110/2] via 10.1.12.2, 00:35:52, GigabitEthernet0/0
O IA    10.2.34.0/24 [110/3] via 10.1.12.2, 00:35:52, GigabitEthernet0/0
O IA 192.168.4.0/23 [110/4] via 10.1.12.2, 00:35:32, GigabitEthernet0/0
```

由以上输出可知，在路由器 R1 的路由表中，原来的 2 条/24 的域间路由汇总成了一条/23 的路由。

```
R2>show ip route | include 110
O IA    10.2.34.0/24 [110/2] via 10.0.23.3, 00:36:24, GigabitEthernet0/1
O       192.168.0.0/24 [110/2] via 10.1.12.1, 00:36:34, GigabitEthernet0/0
O       192.168.1.0/24 [110/2] via 10.1.12.1, 00:36:34, GigabitEthernet0/0
O       192.168.2.0/24 [110/2] via 10.1.12.1, 00:36:34, GigabitEthernet0/0
O       192.168.3.0/24 [110/2] via 10.1.12.1, 00:36:34, GigabitEthernet0/0
O IA 192.168.4.0/23 [110/3] via 10.0.23.3, 00:36:04, GigabitEthernet0/1
```

由以上输出可知，在路由器 R2 的路由表中，原来的 2 条/24 的域间路由汇总成了一条/23 的路由。

```
R3>show ip route | include 110
O IA    10.1.12.0/24 [110/2] via 10.0.23.2, 00:36:29, GigabitEthernet0/1
O IA 192.168.0.0/22 [110/3] via 10.0.23.2, 00:06:24, GigabitEthernet0/1
O       192.168.4.0/24 [110/2] via 10.2.34.4, 00:36:39, GigabitEthernet0/2
O       192.168.5.0/24 [110/2] via 10.2.34.4, 00:36:39, GigabitEthernet0/2
```

由以上输出可知，在路由器 R3 的路由表中，原来的 4 条/24 的域间路由汇总成了 1 条/22 的路由。

```
R4>show ip route | include 110
```

```
O IA    10.0.23.0/24 [110/2] via 10.2.34.3, 00:37:00, GigabitEthernet0/2
O IA    10.1.12.0/24 [110/3] via 10.2.34.3, 00:36:50, GigabitEthernet0/2
O IA 192.168.0.0/22 [110/4] via 10.2.34.3, 00:06:41, GigabitEthernet0/2
```

由以上输出可知，在路由器 R4 的路由表中，原来的 4 条/24 的域间路由汇总成了 1 条/22 的路由。

4.4.3 场景六：配置 OSPFv2 外部路由汇总

通过对 OSPFv2 路由汇总理论知识的学习，相信大家已经对 OSPFv2 路由汇总有了一定的认识。接下来，让我们一起通过具体实验完成 OSPFv2 外部路由汇总配置。OSPFv2 外部路由汇总配置实验拓扑如图 4-7 所示。

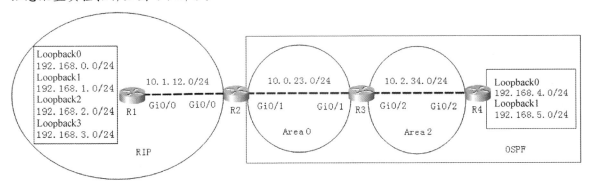

图 4-7　OSPFv2 外部路由汇总配置实验拓扑

任务要求：
- 按图 4-7 所示搭建网络拓扑；
- 按图 4-7 所示为路由器接口配置 IP 地址；
- 按图 4-7 所示配置动态路由协议 RIPv2 和 OSPFv2；
- 按图 4-7 所示配置 RIPv2 和 OSPFv2 的路由重分布使全网互通；
- 配置 OSPFv2 的域间路由汇总和 OSPFv2 外部路由汇总，优化路由表。

配置外部路由汇总命令语法如下（以路由器 R2 为例）：

```
R2(config-)#router ospf 1
R2(config-router)#summary-address 192.168.0.0 255.255.252.0    //配置外部自治系统路由汇总
```

任务实施如下所述。

步骤一：配置路由器接口 IP 地址

路由器 R1、R2、R3 和 R4 的接口配置完全同 4.4.2 节场景五的步骤一。

步骤二：配置动态路由协议 RIPv2 和 OSPFv2

（1）配置路由器 R1

```
R1(config)#router rip
R1(config-router)#version 2
R1(config-router)#network 10.0.0.0
R1(config-router)#network 192.168.0.0
R1(config-router)#network 192.168.1.0
R1(config-router)#network 192.168.2.0
R1(config-router)#network 192.168.3.0
```

（2）配置路由器 R2

```
R2(config)#router rip
R2(config-router)#version 2
R2(config-router)#network 10.0.0.0
R2(config-router)#router ospf 1
R2(config-router)#router-id 2.2.2.2
R2(config-router)#network 10.0.23.0 0.0.0.255 area 0
```

（3）配置路由器 R3

```
R3(config)#router ospf 1
R3(config-router)#router-id 3.3.3.3
R3(config-router)#network 10.0.23.0 0.0.0.255 area 0
R3(config-router)#network 10.2.34.0 0.0.0.255 area 2
```

（4）配置路由器 R4

```
R4(config)#router ospf 1
R4(config-router)#router-id 4.4.4.4
R4(config-router)#network 10.2.34.0 0.0.0.255 area 2
R4(config-router)#network 192.168.4.0 0.0.0.255 area 2
R4(config-router)#network 192.168.5.0 0.0.0.255 area 2
R4(config-router)#interface Loopback 0
R4(config-if)#ip ospf network point-to-point
R4(config-if)#interface Loopback 1
R4(config-if)#ip ospf network point-to-point
```

步骤三：查看路由器的路由表

我们依次查看路由器的路由表发现，在路由器 R1 的路由表中添加了一条 RIP 路由；在路由器 R2 的路由表中添加了所有网段的路由，路由表完整；在路由器 R3 和 R4 的路由表中仅添加了 OSPFv2 的路由，其中 R3 的路由表显示如下。

```
R3>show ip route | begin 10.0.0.0
     10.0.0.0/8 is variably subnetted, 4 subnets, 2 masks
C       10.0.23.0/24 is directly connected, GigabitEthernet0/1
L       10.0.23.3/32 is directly connected, GigabitEthernet0/1
C       10.2.34.0/24 is directly connected, GigabitEthernet0/2
L       10.2.34.3/32 is directly connected, GigabitEthernet0/2
O    192.168.4.0/24 [110/2] via 10.2.34.4, 00:02:03, GigabitEthernet0/2
O    192.168.5.0/24 [110/2] via 10.2.34.4, 00:02:03, GigabitEthernet0/2
```

从以上路由器 R3 的路由表可知，R3 的路由表添加了 2 条 OSPFv2 域内路由。

步骤四：配置 OSPFv2 域间路由汇总和路由重分布

（1）在 ABR 路由器 R3 上配置 OSPFv2 域间路由汇总

```
R3(config)#router ospf 1
R3(config-router)#area 2 range 192.168.4.0 255.255.254.0
```

（2）在 ABR 路由器 R2 上配置 RIPv2 和 OSPFv2 路由重分布

```
R2(config)#router ospf 1
R2(config-router)#redistribute rip subnets
R2(config-router)#router rip
R2(config-router)#redistribute ospf 1 metric 11    //RIP 规定 16 跳不可达，这里度量设置为 11
```

步骤五：查看路由器路由表的变化

```
R1>show ip route | include 11
R       10.2.34.0/24 [120/11] via 10.1.12.2, 00:00:18, GigabitEthernet0/0
R       192.168.4.0/23 [120/11] via 10.1.12.2, 00:00:18, GigabitEthernet0/0
```

由以上输出可知，在路由器 R1 的路由表增加了 2 条度量为 11 的 RIP 路由，其中一条是汇总路由，该汇总路由也同时出现在路由器 R2 的路由表中。路由器 R3 和 R4 的路由表都添加了 OSPFv2 的外部路由（RIP 路由域），其中 R3 的路由表显示如下。

```
R3>show ip route | include O E
O E2    10.1.12.0/24 [110/20] via 10.0.23.2, 01:00:30, GigabitEthernet0/1
```

```
O E2 192.168.0.0/24 [110/20] via 10.0.23.2, 01:00:30, GigabitEthernet0/1
O E2 192.168.1.0/24 [110/20] via 10.0.23.2, 01:00:30, GigabitEthernet0/1
O E2 192.168.2.0/24 [110/20] via 10.0.23.2, 01:00:30, GigabitEthernet0/1
O E2 192.168.3.0/24 [110/20] via 10.0.23.2, 01:00:30, GigabitEthernet0/1
```

步骤六：在 ASBR 上配置外部路由汇总

```
R2(config)#router ospf 1
R2(config-router)#summary-address 192.168.0.0 255.255.252.0
```

注意：当前 Cisco Packet Tracer（8.0 及以前版本）尚不支持 **summary-address** 外部路由汇总命令，本实验将 Packet Tracer（PT）与真实实验平台相结合，PT 支持步骤一～五，步骤六～七切换到真实平台。

步骤七：查看路由器路由表的变化

```
R3>show ip route | include O E
O E2      10.1.12.0/24 [110/20] via 10.0.23.2, 01:00:30, GigabitEthernet0/1
O E2 192.168.0.0/22 [110/20] via 10.0.23.2, 01:00:30, GigabitEthernet0/1
```

在以上路由器 R3 的路由表中，192.168.0.0/22 为外部汇总路由。

```
R4>show ip route | include O E
O E2      10.1.12.0/24 [110/20] via 10.2.34.3, 01:06:36, GigabitEthernet0/2
O E2 192.168.0.0/22 [110/20] via 10.2.34.3, 01:06:36, GigabitEthernet0/2
```

在以上路由器 R4 的路由表中，192.168.0.0/22 为外部汇总路由。

4.5 完成 OSPFv2 特殊区域配置

4.5.1 认识 OSPFv2 特殊区域

OSPFv2 区域采用两级结构，具体类型如下所述。

① 标准区域：可以接收链路更新信息，以及相同区域的路由、区域间路由和外部 AS 的路由信息。标准区域通常与区域 0 连接。

② 主干区域：连接各个标准区域的中心实体，标准区域都要连接到该区域交换路由信息。主干区域也叫区域 0。

OSPFv2 采用分层区域结构来构建大规模网络，为进一步优化网络，提升网络性能，OSPFv2 特别设计了以下 4 种特殊区域作为对区域设计的补充。

1. 末节区域

末节区域（Stub Area）：代表着"网络的尽头"。通常，末节区域只有单一的区域出口，进入或离开末节区域的数据包只能通过 ABR 转发，该区域不接收 AS 以外的路由信息；末节区域内部的路由器不能作为 ASBR。由于末节区域没有去往外部网络的路由，所以 ABR 会自动向末节区域发送一条指向自己的默认路由（0.0.0.0）。末节区域的内部路由器将使用一条默认路由替代所有外部路由条目，从而缩小路由表的规模。

2. 完全末节区域

完全末节区域（Totally Stub Area）：又被称为绝对末节区域，属于 Cisco 专有特殊区域，该区域既不接收 AS 外部的路由信息，也不接收 AS 内其他区域的汇总路由信息。如果路由器需要路由数据包至本区域以外的区域，那么它将使用默认路由（0.0.0.0），该默认路由也是由 ABR 向完全末节区域自动注入进来的一条指向自己的开销为 1 的全零路由。该区域同样也不存在 ASBR。在完全末节区域的路由器的路由表中有域内路由和默认路由，但不存在域间路由和外部路由。

在配置末节区域的命令后增加关键字 no-summary 即可完成完全末节区域配置，表示不需要通告汇总。对完全末节区域的配置仅需在 ABR 上完成即可，无须对完全末节区域内部的路由器进行相应配置。

3. 次末节区域

次末节区域（Not-So-Stubby Area，NSSA）：该区域与末节区域类似，是思科专有属性，是对末节区域特性的扩展，同时也是一种与传统末节区域相违背的区域设计方式。传统末节区域不接收外部路由，也不允许该区域内存在 ASBR；而次末节区域允许存在 ASBR，允许 ASBR 引入外部路由。换言之，次末节区域允许连接外部网络，例如，可以连接运行 RIP 和 OSPF 的网络等。

在 NSSA 内部，对每台路由器都必须配置 NSSA，这样它们才能建立邻居关系并进一步交换路由信息，否则无法建立邻居关系。

在 NSSA 中，允许 7 类 LSA 存在，由 ABR 负责将 7 类 LSA 转换成 5 类 LSA，同时阻止 4 类和 5 类 LSA。在 NSSA 中，因为允许将外部路由重分布进 OSPFv2 进程，所以 ABR 不会自动向 NSSA 内部自动注入一条指向自己的默认路由。但是在 ABR 上，可以通过手工方式注入一条默认路由，特别值得注意的是，在本 NSSA 中，此时的 ABR 同时又是 ASBR。

4. 完全次末节区域

完全次末节区域（Totally Not-So-Stubby Area，Totally NSSA）：又被称为绝对次末节区域，该区域也属于 Cisco 专有的特殊区域，是对次末节区域特性的扩展。Totally NSSA 的 ABR 允许进行域间路由，也就是允许发送 3 类 LSA 来实现域间路由。

在 Cisco 路由器上配置 Totally NSSA 时也只需要在 ABR 上进行设置，只要在 NSSA 命令后增加关键字 no-summary 即可将 NSSA 转换成一个 Totally NSSA。关键字 no-summary 的意思是不通告汇总，即阻止 3 类和 4 类 LSA 的汇总路由扩散至 NSSA，从而使 NSSA 成为 Totally NSSA。

4.5.2　场景七：配置 OSPFv2 特殊区域

通过前面内容的学习，相信大家已经对 OSPFv2 的几种特殊区域有了初步认知，接下来，让我们一起通过具体实验完成特殊区域配置。配置 OSPFv2 特殊区域实验拓扑如图 4-8 所示。

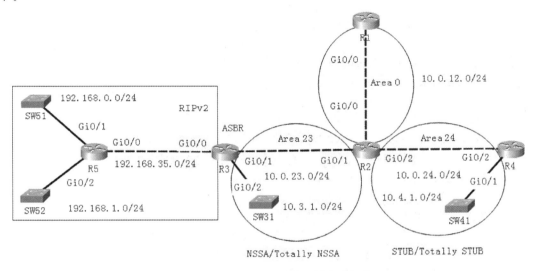

图 4-8　配置 OSPFv2 特殊区域实验拓扑

任务需求：
- 按图 4-8 所示搭建拓扑；
- 按图 4-8 所示为设备配置 IP 地址；
- 按图 4-8 所示配置 OSPFv2 和 RIPv2；
- 配置 OSPFv2 特殊区域使得全网连通。

OSPFv2 配置末节区域命令如下：

```
Router(config)#router ospf 1
Router(config-router)#area 1 stub                    //将 area 1 配置为末节区域
Router(config-router)#area 1 stub no-summary         //将 area 1 配置为完全末节区域
Router(config-router)#area 1 nssa                    //将 area 1 配置为次末节区域
Router(config-router)#area 1 nssa no-summary         //将 area 1 配置为完全次末节区域
```

任务实施如下所述。

步骤一：配置路由器接口 IP 地址

（1）配置路由器 R1

```
R1(config)#interface GigabitEthernet 0/0
R1(config-if)#ip address 10.0.12.1 255.255.255.0
R1(config-if)#no shutdown
```

（2）配置路由器 R2

```
R2(config)#interface GigabitEthernet 0/0
R2(config-if)#ip address 10.0.12.2 255.255.255.0
R2(config-if)#no shutdown
R2(config-if)#interface GigabitEthernet 0/1
R2(config-if)#ip address 10.0.23.2 255.255.255.0
R2(config-if)#no shutdown
R2(config-if)#interface GigabitEthernet 0/2
R2(config-if)#ip address 10.0.24.2 255.255.255.0
R2(config-if)#no shutdown
```

（3）配置路由器 R3

```
R3(config-if)#interface GigabitEthernet 0/0
R3(config-if)#ip address 192.168.35.3 255.255.255.0
R3(config-if)#no shutdown
R3(config)#interface GigabitEthernet 0/1
R3(config-if)#ip address 10.0.23.3 255.255.255.0
R3(config-if)#no shutdown
R3(config)#interface GigabitEthernet 0/2
R3(config-if)#ip address 10.3.1.254 255.255.255.0
R3(config-if)#no shutdown
```

（4）配置路由器 R4

```
R4(config)#interface GigabitEthernet 0/1
R4(config-if)#ip address 10.4.1.254 255.255.255.0
R4(config-if)#no shutdown
R4(config)#interface GigabitEthernet 0/2
R4(config-if)#ip address 10.0.24.4 255.255.255.0
```

R4(config-if)#**no shutdown**

（5）配置路由器 R5

R5(config)#**interface GigabitEthernet 0/1**
R5(config-if)#**ip address 192.168.0.254 255.255.255.0**
R5(config-if)#**no shutdown**
R5(config-if)#**interface GigabitEthernet 0/2**
R5(config-if)#**ip address 192.168.1.254 255.255.255.0**
R5(config-if)#**no shutdown**
R5(config-if)#**interface GigabitEthernet 0/0**
R5(config-if)#**ip address 192.168.35.5 255.255.255.0**

（6）配置交换机 IP 地址为 X.X.X.100/24，网关为 X.X.X.254

SW51(config)#**interface vlan 1**
SW51(config-if)#**ip address 192.168.0.100 255.255.255.0**
SW51(config-if)#**no shutdown**
SW51(config-if)#**exit**
SW51(config)#**ip default-gateway 192.168.0.254**

请参照交换机 SW51 的配置方式完成交换机 SW52、SW31 和 SW41 的 IP 地址和网关配置。

步骤二：配置路由协议 OSPFv2

（1）配置路由器 R1

R1(config)#**router ospf 1**
R1(config-router)#**router-id 1.1.1.1**
R1(config-router)#**network 10.0.12.0 0.0.0.255 area 0**

（2）配置路由器 R2

R2(config)#**router ospf 1**
R2(config-router)#**router-id 2.2.2.2**
R2(config-router)#**network 10.0.12.0 0.0.0.255 area 0**
R2(config-router)#**network 10.0.23.0 0.0.0.255 area 23**
R2(config-router)#**network 10.0.24.0 0.0.0.255 area 24**

（3）配置路由器 R3

R3(config)#**router ospf 1**
R3(config-router)#**router-id 3.3.3.3**
R3(config-router)#**network 10.0.23.0 0.0.0.255 area 23**

R3(config-router)#**network 10.3.1.0 0.0.0.255 area 23**

（4）配置路由器 R4

R4(config)#**router ospf 1**
R4(config-router)#**router-id 4.4.4.4**
R4(config-router)#**network 10.0.24.0 0.0.0.255 area 24**
R4(config-router)#**network 10.4.1.0 0.0.0.255 area 24**

步骤三：配置路由协议 RIPv2

（1）配置路由器 R3

R3(config)#**router rip**
R3(config-router)#**version 2**
R3(config-router)#**no auto-summary**　　//关闭自动汇总，避免步骤四中 OSPFv2 路由域路由自动汇总
R3(config-router)#**network 192.168.35.0**

（2）配置路由器 R5

R5(config)#**router rip**
R5(config-router)#**version 2**
R3(config-router)#**no auto-summary**　　//关闭自动汇总，避免步骤四中 OSPFv2 路由域路由自动汇总
R5(config-router)#**network 192.168.0.0**
R5(config-router)#**network 192.168.1.0**
R5(config-router)#**network 192.168.35.0**

步骤四：配置路由注入

（1）配置路由器 R3 实现 OSPFv2 和 RIPv2 的路由注入

R3(config)#**router rip**
R3(config-router)#**redistribute ospf 1 metric 11**　　//RIPv2 规定 16 跳不可达，此处将度量设置为 11
R3(config-router)#**router ospf 1**
R3(config-router)#**redistribute rip**

（2）查看路由器 R5 的路由表

R5#**show ip route | begin Gateway**
Gateway of last resort is not set
　　　10.0.0.0/24 is subnetted, 5 subnets
R　　　10.0.12.0/24 [120/11] via 192.168.35.3, 00:00:01, GigabitEthernet0/0
R　　　10.0.23.0/24 [120/11] via 192.168.35.3, 00:00:01, GigabitEthernet0/0

R	10.0.24.0/24 [120/11] via 192.168.35.3, 00:00:01, GigabitEthernet0/0	
R	10.3.1.0/24 [120/11] via 192.168.35.3, 00:00:01, GigabitEthernet0/0	
R	10.4.1.0/24 [120/11] via 192.168.35.3, 00:00:01, GigabitEthernet0/0	
	192.168.0.0/24 is variably subnetted, 2 subnets, 2 masks	
C	192.168.0.0/24 is directly connected, GigabitEthernet0/1	
L	192.168.0.254/32 is directly connected, GigabitEthernet0/1	
	192.168.1.0/24 is variably subnetted, 2 subnets, 2 masks	
C	192.168.1.0/24 is directly connected, GigabitEthernet0/2	
L	192.168.1.254/32 is directly connected, GigabitEthernet0/2	
	192.168.35.0/24 is variably subnetted, 2 subnets, 2 masks	
C	192.168.35.0/24 is directly connected, GigabitEthernet0/0	
L	192.168.35.5/32 is directly connected, GigabitEthernet0/0	

（3）测试网络连通性，全网互通

```
SW51#traceroute 10.4.1.100
Tracing the route to 10.4.1.100
  1    192.168.0.254    0 msec    0 msec    0 msec    //R5
  2    192.168.35.3     0 msec    0 msec    1 msec    //R3
  3    10.0.23.2        0 msec    0 msec    0 msec    //R2
  4    10.0.24.4        0 msec    0 msec    0 msec    //R4
  5    10.4.1.100       11 msec   11 msec   11 msec   //SW41
```

以上输出表明，192.168.0.0/24 网段可以与 10.4.1.0/24 网段互访。

```
SW52>traceroute 10.3.1.100
Tracing the route to 10.3.1.100
  1    192.168.1.254    0 msec    0 msec    0 msec    //R5
  2    192.168.35.3     0 msec    0 msec    0 msec    //R3
  3    10.3.1.100       0 msec    0 msec    0 msec    //SW31
```

以上输出表明，192.168.1.0/24 网段可以与 10.3.1.0/24 网段互访。

步骤五：配置 Area 24 为 Stub 区域

末节区域（Stub 区域）会阻止 4 类和 5 类 LSA 传入区域内，由 ABR 向区域内的路由器下发一条指向自己的默认路由，从而优化路由表，提高网络性能。

（1）在配置 Stub 区域前查看路由器 R4 的拓扑表和路由表

```
R4>show ip ospf database
    OSPF Router with ID (4.4.4.4) (Process ID 1)
```

Router Link States (Area 24) //1 类 LSA

Link ID	ADV Router	Age	Seq#	Checksum	Link count
4.4.4.4	4.4.4.4	565	0x80000004	0x008d38	2
2.2.2.2	2.2.2.2	565	0x80000002	0x005f99	1

Net Link States (Area 24) //2 类 LSA

Link ID	ADV Router	Age	Seq#	Checksum
10.0.24.4	4.4.4.4	565	0x80000001	0x006127

Summary Net Link States (Area 24)//3 类 LSA

Link ID	ADV Router	Age	Seq#	Checksum
10.0.12.0	2.2.2.2	560	0x80000001	0x0050ef
10.0.23.0	2.2.2.2	560	0x80000002	0x00d45f
10.3.1.0	2.2.2.2	560	0x80000004	0x00a99a

Summary ASB Link States (Area 24)//4 类 LSA

Link ID	ADV Router	Age	Seq#	Checksum
3.3.3.3	2.2.2.2	560	0x80000003	0x00ba8c

Type-5 AS External Link States //5 类 LSA

Link ID	ADV Router	Age	Seq#	Checksum	Tag
192.168.35.0	3.3.3.3	609	0x80000001	0x003bec	0
192.168.0.0	3.3.3.3	609	0x80000001	0x00bd8d	0
192.168.1.0	3.3.3.3	609	0x80000001	0x00b297	0

以上输出表明，路由器 R4 的拓扑表中存在 5 种 LSA，其中 3 类 LSA 和 5 类 LSA 均有 3 个网段。

```
R4>show ip route | begin Gateway
Gateway of last resort is not set
     10.0.0.0/8 is variably subnetted, 7 subnets, 2 masks
O IA    10.0.12.0/24 [110/2] via 10.0.24.2, 00:15:58, GigabitEthernet0/2
O IA    10.0.23.0/24 [110/2] via 10.0.24.2, 00:15:58, GigabitEthernet0/2
C       10.0.24.0/24 is directly connected, GigabitEthernet0/2
L       10.0.24.4/32 is directly connected, GigabitEthernet0/2
O IA    10.3.1.0/24 [110/3] via 10.0.24.2, 00:15:58, GigabitEthernet0/2
C       10.4.1.0/24 is directly connected, GigabitEthernet0/1
L       10.4.1.254/32 is directly connected, GigabitEthernet0/1
O E2 192.168.0.0/24 [110/20] via 10.0.24.2, 00:15:58, GigabitEthernet0/2
O E2 192.168.1.0/24 [110/20] via 10.0.24.2, 00:15:58, GigabitEthernet0/2
O E2 192.168.35.0/24 [110/20] via 10.0.24.2, 00:15:58, GigabitEthernet0/2
```

以上输出表明，路由器 R4 的路由表中有 3 条域间路由（O IA）和 3 条外部路由（O E2）。

（2）配置 Area 24 为 Stub 区域

在 ABR R2 上配置 Stub 区域：

  R2(config)#**router ospf 1**
  R2(config-router)#**area 24 stub**

在路由器 R4 上配置 Stub 区域：

  R4(config)#**router ospf 1**
  R4(config-router)#**area 24 stub**

（3）在配置 Stub 区域后查看路由器 R4 的拓扑表和路由表

```
R4>show ip ospf database
          OSPF Router with ID (4.4.4.4) (Process ID 1)
            Router Link States (Area 24)      //1 类 LSA
Link ID         ADV Router        Age        Seq#          Checksum Link count
4.4.4.4         4.4.4.4           695        0x80000004    0x008d38  2
2.2.2.2         2.2.2.2           695        0x80000004    0x005b9b  1
            Net Link States (Area 24)         //2 类 LSA
Link ID         ADV Router        Age        Seq#          Checksum
10.0.24.4       4.4.4.4           695        0x80000001    0x00ecba
            Summary Net Link States (Area 24)  //3 类 LSA
Link ID         ADV Router        Age        Seq#          Checksum
0.0.0.0         2.2.2.2           685        0x80000003    0x005301
10.0.12.0       2.2.2.2           685        0x80000001    0x0050ef
10.0.23.0       2.2.2.2           685        0x80000002    0x00d45f
10.3.1.0        2.2.2.2           680        0x80000004    0x00a99a
```

以上输出表明，当配置了 Stub 区域后，路由器 R4 的拓扑表中只有 3 种 LSA，4 类和 5 类 LSA 消失。其中 3 类 LSA 中增加了一条 0.0.0.0 的路由。

```
R4>show ip route | begin Gateway
Gateway of last resort is 10.0.24.2 to network 0.0.0.0
      10.0.0.0/8 is variably subnetted, 7 subnets, 2 masks
O IA     10.0.12.0/24 [110/2] via 10.0.24.2, 00:11:20, GigabitEthernet0/2
O IA     10.0.23.0/24 [110/2] via 10.0.24.2, 00:11:20, GigabitEthernet0/2
C        10.0.24.0/24 is directly connected, GigabitEthernet0/2
L        10.0.24.4/32 is directly connected, GigabitEthernet0/2
O IA     10.3.1.0/24 [110/3] via 10.0.24.2, 00:11:20, GigabitEthernet0/2
C        10.4.1.0/24 is directly connected, GigabitEthernet0/1
L        10.4.1.254/32 is directly connected, GigabitEthernet0/1
```

```
O*IA 0.0.0.0/0 [110/2] via 10.0.24.2, 00:11:20, GigabitEthernet0/2
```

由以上输出可知，配置 Stub 区域后，路由器 R4 的路由表中原有的标记为"O E2"的外部路由 192.168.0.0/24、192.168.1.0/24 和 192.168.35.0/24 消失了，它们被 ABR R2 下发的带"O*IA"标记的默认路由 0.0.0.0/0 取代，路由器 R4 如果要访问外部 RIP 路由域，通过默认路由可达。

（4）Stub 区域访问 RIP 路由域测试

```
SW41>traceroute 192.168.0.100
Tracing the route to 192.168.0.100
  1   10.4.1.254      0 msec    0 msec    0 msec
  2   10.0.24.2       0 msec    0 msec    0 msec
  3   10.0.23.3       0 msec    0 msec    0 msec
  4   192.168.35.5    0 msec    0 msec    0 msec
  5   192.168.0.100   11 msec   0 msec    11 msec
```

以上输出显示，SW41 可以访问 SW51，即 Stub 区域可以访问外部 RIP 路由域。

步骤六：配置 Area 24 为 Totally Stub 区域

完全末节区域（Totally Stub）会在 Stub 区域阻止 4 类和 5 类 LSA 传入区域的基础上，进一步阻止 3 类 LSA 传入，从而减少 LSA，优化路由表，提升网络性能。

（1）在 ABR 上配置 Totally Stub 区域

```
R2(config)#router ospf 1
R2(config-router)#area 24 stub no-summary
```

Totally Stub 区域只需要在 ABR 上配置，这样可以阻止 3 类 LSA 下发到该区域。

（2）查看 Totally Stub 区域路由器 R4 的拓扑表和路由表

```
R4>show ip ospf database
        OSPF Router with ID (4.4.4.4) (Process ID 1)
              Router Link States (Area 24)
Link ID         ADV Router       Age        Seq#          Checksum Link count
4.4.4.4         4.4.4.4          1054       0x80000004    0x008d38  2
2.2.2.2         2.2.2.2          1054       0x80000004    0x005b9b  1
              Net Link States (Area 24)
Link ID         ADV Router       Age        Seq#          Checksum
10.0.24.4       4.4.4.4          1054       0x80000001    0x00ecba
              Summary Net Link States (Area 24)
Link ID         ADV Router       Age        Seq#          Checksum
0.0.0.0         2.2.2.2          1044       0x80000003    0x005301
```

以上输出表明，配置了 Totally Stub 区域后，路由器 R4 拓扑表中 3 类 LSA 的记录 10.0.12.0、10.0.23.0 和 10.3.1.0 消失，即 3 类 LSA 被屏蔽，但 ABR R2 生成的 0.0.0.0 默认路由依然存在。

```
R4>show ip route | begin Gateway
Gateway of last resort is 10.0.24.2 to network 0.0.0.0

      10.0.0.0/8 is variably subnetted, 4 subnets, 2 masks
C        10.0.24.0/24 is directly connected, GigabitEthernet0/2
L        10.0.24.4/32 is directly connected, GigabitEthernet0/2
C        10.4.1.0/24 is directly connected, GigabitEthernet0/1
L        10.4.1.254/32 is directly connected, GigabitEthernet0/1
O*IA 0.0.0.0/0 [110/2] via 10.0.24.2, 00:17:23, GigabitEthernet0/2
```

以上输出表明，配置 Totally Stub 区域后，路由器 R4 路由表中原有的标记为"**O IA**"的域间路由 10.0.12.0/24、10.0.23.0/24 和 10.3.1.0/24 消失，路由器 R4 如果访问外部 RIP 路由域及 OSPF 其他区域，通过默认路由可达。

（3）Totally Stub 区域访问 RIPv2 路由域和 OSPFv2 其他区域

```
SW41>ping 192.168.1.100
Sending 5, 100-byte ICMP Echos to 192.168.1.100, timeout is 2 seconds:
!!!!!
Success rate is 100 percent (5/5), round-trip min/avg/max = 0/2/10 ms
```

以上输出表明，SW41 可以访问 SW52，即可以访问外部 AS。

```
SW41>ping 10.3.1.100
Sending 5, 100-byte ICMP Echos to 10.3.1.100, timeout is 2 seconds:
!!!!!
Success rate is 100 percent (5/5), round-trip min/avg/max = 0/0/0 ms
```

以上输出表明，SW41 可以访问 SW31，即可以访问 Area 23。

```
SW41>ping 10.0.12.1
Sending 5, 100-byte ICMP Echos to 10.0.12.1, timeout is 2 seconds:
!!!!!
Success rate is 100 percent (5/5), round-trip min/avg/max = 0/0/0 ms
```

以上输出表明，SW41 可以访问 R1，即可以访问 Area 0。
Totally Stub 区域屏蔽了 3 类、4 类和 5 类 LSA，极大程度优化了路由表，提升网络性能。

步骤七：配置 Area 23 为 NSSA 区域

（1）在配置 NSSA 区域前查看 Area 23 路由器的拓扑表和路由表

查看路由器 R3 的拓扑表，我们可以看到该拓扑表中有 1 类、2 类、3 类和 5 类共 4 种 LSA。

同理，路由器 R2 在 Area 23 中的拓扑表与路由器 R3 的拓扑表一致，LSA 类型一致，其 5 类 LSA 如下所示。

	Type-5 AS External Link States				
Link ID	ADV Router	Age	Seq#	Checksum	Tag
192.168.35.0	3.3.3.3	450	0x80000001	0x003bec	0
192.168.0.0	3.3.3.3	450	0x80000001	0x00bd8d	0
192.168.1.0	3.3.3.3	450	0x80000001	0x00b297	0

以上路由器 R2 拓扑表部分输出表明，路由器 R3（3.3.3.3）是 ASBR，也就是 5 类 LSA 由其产生。R2 路由表部分输出如下。

```
R2>show ip route | include 110
O       10.3.1.0/24 [110/2] via 10.0.23.3, 01:56:10, GigabitEthernet0/1
O       10.4.1.0/24 [110/2] via 10.0.24.4, 01:56:00, GigabitEthernet0/2
O E2 192.168.0.0/24 [110/20] via 10.0.23.3, 01:56:10, GigabitEthernet0/1
O E2 192.168.1.0/24 [110/20] via 10.0.23.3, 01:56:10, GigabitEthernet0/1
O E2 192.168.35.0/24 [110/20] via 10.0.23.3, 01:56:10, GigabitEthernet0/1
```

以上输出表明，路由器 R2 通过下一跳路由器 R3（10.0.23.3）学习到了 3 条外部路由，标记为"O E2"。

（2）配置 Area 23 为 NSSA 区域

在 ABR R2 上配置 NSSA 区域：

```
R2(config)#router ospf 1
R2(config-router)#area 23 nssa
```

在路由器 R3 上配置 NSSA 区域：

```
R3(config)#router ospf 1
R3(config-router)#area 23 nssa
```

（3）在配置 NSSA 区域后查看 Area 23 路由器的拓扑表和路由表

查看路由器 R3 的拓扑表，我们可以看到该拓扑表中有 4 种 LSA，5 类 LSA 消失了，出现 7 类 LSA。

我们接着查看路由器 R2 在 Area 23 的拓扑表，其 7 类 LSA 显示如下。

	Type-7 AS External Link States (Area 23)				
Link ID	ADV Router	Age	Seq#	Checksum	Tag
192.168.35.0	3.3.3.3	472	0x80000012	0x0066a4	0
192.168.0.0	3.3.3.3	418	0x80000013	0x00ab5d	0
192.168.1.0	3.3.3.3	418	0x80000014	0x009e68	0

此时，我们发现路由器 R2 的拓扑表中依然存在 5 类 LSA，具体如下所示。

Type-5 AS External Link States					
Link ID	ADV Router	Age	Seq#	Checksum	Tag
192.168.35.0	2.2.2.2	875	0x80000001	0x0059d2	0
192.168.0.0	2.2.2.2	875	0x80000002	0x0074b8	0
192.168.1.0	2.2.2.2	875	0x80000003	0x0067c3	0

以上表明，路由器 R2 的拓扑表中有 5 类 LSA，且由 R2 生成，即 ABR R2 负责将 7 类 LSA 转换成了 5 类 LSA，因为 Area 24 屏蔽了 5 类 LSA，因此我们接着查看路由器 R1 的拓扑表，发现 5 类 LSA 已传至路由器 R1。7 类 LSA 只能在 NSSA 区域内传输，必须由 ABR 将其转换成 5 类 LSA 后才可传输至其他区域。

接下来，我们来查看路由器 R2 的路由表。

```
R2>show ip route | include 110
O        10.3.1.0/24 [110/2] via 10.0.23.3, 00:19:51, GigabitEthernet0/1
O        10.4.1.0/24 [110/2] via 10.0.24.4, 00:20:01, GigabitEthernet0/2
O N2 192.168.0.0/24 [110/20] via 10.0.23.3, 00:19:51, GigabitEthernet0/1
O N2 192.168.1.0/24 [110/20] via 10.0.23.3, 00:19:51, GigabitEthernet0/1
O N2 192.168.35.0/24 [110/20] via 10.0.23.3, 00:19:51, GigabitEthernet0/1
```

以上输出表明，原来外部标记为"O E2"的路由，现标记已更改为"O N2"。即 NSSA 区域配置成功。请依次进行网段连通性测试，全网可以互通。

步骤八：配置 Area 23 为 Totally NSSA 区域

完全次末节区域（Totally NSSA）会在 NSSA 区域基础上阻止 3 类 LSA 传入，从而减少 LSA，优化路由表，提升网络性能。

（1）在配置 Totally NSSA 区域前查看路由器 R3 的拓扑表和路由表

Summary Net Link States (Area 23)				
Link ID	ADV Router	Age	Seq#	Checksum
10.0.12.0	2.2.2.2	427	0x8000001c	0x001a0b
10.0.24.0	2.2.2.2	427	0x8000001d	0x009384
10.4.1.0	2.2.2.2	427	0x8000001e	0x0069bf

以上路由器 R3 拓扑表的部分输出表明，路由器 R3 拓扑表中存在由 ABR R2 发送的 3 条 3 类 LSA。

```
R3>show ip route | include 110
O IA     10.0.12.0/24 [110/2] via 10.0.23.2, 00:06:36, GigabitEthernet0/1
O IA     10.0.24.0/24 [110/2] via 10.0.23.2, 00:06:36, GigabitEthernet0/1
```

    O IA 10.4.1.0/24 [110/3] via 10.0.23.2, 00:06:36, GigabitEthernet0/

以上输出表明，路由器 R3 的路由表中存在 3 条域间路由，分别来自 Area 0 和 Area 24。

（2）在 ABR 上配置 Area 23 为 Totally NSSA 区域

   R2(config)#**router ospf 1**
   R2(config-router)#**area 23 nssa no-summary**

Totally NSSA 区域只需要在 ABR 上配置，这样可以阻止 3 类 LSA 下发给该区域的路由器。

（3）配置 Totally NSSA 区域后，查看路由器 R3 拓扑表的变化

       Summary Net Link States (Area 23)
   Link ID ADV Router Age   Seq#   Checksum
   0.0.0.0   2.2.2.2   325   0x80000016 0x002d14thernet0/1

以上输出表明，路由器 R3 拓扑表中原有的 3 类 LSA 记录 10.0.12.0、10.0.24.0 和 10.4.1.0 消失，即 Area 23 的 3 类 LSA 被屏蔽，被由 ABR R2 生成的 0.0.0.0 默认路由代替。

（4）在配置 Totally NSSA 区域后查看路由器 R3 路由表的变化

   R3>**show ip route | include 110**
   O*IA 0.0.0.0/0 [110/2] via 10.0.23.2, 00:02:13, GigabitEthernet0/1

以上输出表明，配置 Totally NSSA 后，路由器 R3 的路由表中标记为"**O IA**"的域间路由 10.0.12.0/24、10.0.24.0/24 和 10.4.1.0/24 消失，路由器 R3 访问其他区域，可通过路由器 R2 下发的默认路由转发。

（5）在配置 Totally NSSA 区域后测试网络的连通性

   SW51>**traceroute 10.4.1.100**
   Tracing the route to 10.4.1.100

1	192.168.0.254	0 msec	0 msec	0 msec
2	192.168.35.3	0 msec	0 msec	0 msec
3	10.0.23.2	0 msec	0 msec	0 msec
4	10.0.24.4	0 msec	0 msec	0 msec
5	10.4.1.100	0 msec	0 msec	0 msec

以上输出表明，配置 Totally NSSA 后，网络通信正常。

  在本场景中，我们通过配置 Stub、Totally Stub、NSSA 和 Totally NSSA 四种特殊区域实验，了解了 OSPFv2 特殊区域如何通过 ABR 下发默认路由，屏蔽相应 LSA 来优化路由表，进而提升网络性能。

4.6 挑战练习

4.6.1 挑战练习一

挑战要求：挑战练习一实验拓扑如图 4-9 所示，该拓扑由 10 台路由器和 1 台三层交换机互连而成。请根据本章所学知识，完成 OSPFv2 相关配置，确保全网互通。

请读者根据任务要求独立完成该挑战练习。

任务要求：

- 请按图 4-9 所示，搭建网络拓扑；
- 请用 A 类私有地址规划所有网段；
- Area 2 中所有网段 IP 地址要求连续分配；
- 配置所有设备的主机名及接口 IP 地址；
- 配置所有三层设备环回地址作为路由器 ID；
- 配置多区域 OSPFv2，确保全网实现互通；
- 干预 DR 选举，使 Area 1 中的 R11 为 DR，R12 不能参与选举；
- 配置 OSPFv2 认证，Area 0 中所有路由器采用基于接口的 MD5 认证；
- 配置 Area 2 的域间路由汇总，优化路由表，提升网络性能。

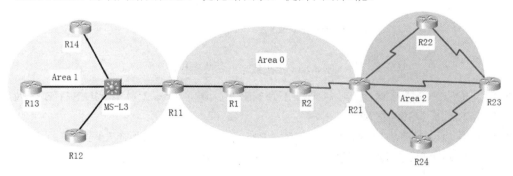

图 4-9 挑战练习一实验拓扑

4.6.2 挑战练习二

挑战要求：挑战练习二实验拓扑如图 4-10 所示，该拓扑由 12 台路由器互连而成。请根据本章所学知识，完成 OSPFv2 相关配置，确保全网互通。

请读者根据任务要求独立完成该挑战练习。

任务要求：

- 请按图 4-10 所示，搭建网络拓扑；

- 请用 A 类私有地址规划路由器 R1～R11 的网络地址，R1 至 ISP 的网段请用公网 IP 地址；
- 配置路由器 R1～R11 环回接口地址为 x.x.x.x/32，例如，R11 的环回地址为 11.11.11.11/32；
- 请配置多区域、多进程 OSPFv2，其中路由器 R3 同时运行 2 个 OSPFv2 进程，与路由器 R2 相连的路由器的接口运行进程 100，与路由器 R4 相连的路由器的接口运行进程 200；
- 配置 OSPFv2 认证，在 Area 0 中配置 MD5 认证；

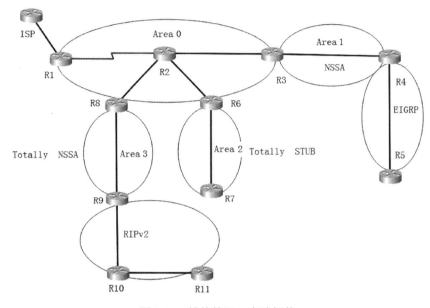

图 4-10 挑战练习二实验拓扑

- 配置 OSPFv2 特殊区域，其中 Area 1 为 NSSA 区域，Area 2 为 Totally STUB 区域，Area 3 为 Totally NSSA 区域；
- 配置其他动态路由协议，路由器 R9～R11 运行 RIPv2，路由器 R4 和 R5 运行 EIGRP；
- 在路由器 R1 上配置静态默认路由，要求在 OSPFv2 网络中传播，在相应路由器上配置路由重分布；
- 除 ISP 外，要求全网段能够相互 ping 通，要求可对全网通过环回接口进行远程管理。

4.7　本章小结

本章内容到此结束。本章主要内容包括 OSPFv2 虚链路配置、认证配置、外部路由注入配置、路由汇总配置以及特殊区域配置。通过本章的学习，读者可以深入理解 OSPFv2 的高级特性，本章中的 7 个应用场景和 2 个挑战练习，有助于读者在实际网络中灵活应用 OSPFv2。

第5章 >>>

学习 ACL 技术

本章要点：

- 认识 ACL 技术
- 配置标准 ACL
- 配置扩展 ACL
- 拓展 ACL 应用
- 挑战练习
- 本章小结

本章通过介绍 ACL（访问控制列表）相关概念，使读者能够理解 ACL 的包过滤规则。其中，5.1 节介绍 ACL 定义、通配符掩码、常用端口号、ACL 应用原则以及应用范围；5.2 节介绍标准 ACL 的特点、标准标号 ACL 和标准命名 ACL 的配置；5.3 节介绍扩展 ACL 的特点、扩展编号 ACL 和扩展命名 ACL 的配置；5.4 节介绍限制 VTY 访问和应用于 QoS 的拓展 ACL。本章设计了 6 个应用场景和 2 个挑战练习，使读者能够熟练应用 ACL 技术，增强网络的安全性。

5.1 认识 ACL 技术

5.1.1 认识 ACL 定义

ACL（Access Control List，访问控制列表）是一条或多条允许或拒绝指令的集合。ACL 本质是一种报文过滤器，通过定义相关策略来匹配报文。ACL 可以对网络中的报文精准匹配，从而实现对网络行为的管控。ACL 是一种控制网络访问的有利的安全工具。

5.1.2 学习通配符掩码

路由器使用通配符掩码与源 IP 地址或者目标 IP 地址一起来分辨所匹配的 IP 地址范围。在访问控制列表中，通配符掩码位为 1，表示忽略（不检查）IP 地址中的对应位；通配符掩码位为 0，表示必须精确匹配（检查）IP 地址中的对应位。通配符掩码应用如表 5-1 所示。

表 5-1 通配符掩码应用

匹配项	匹配 IP 地址	通配符掩码
一台主机	192.168.1.1/32	0.0.0.0
一个主网	192.168.1.0/24	0.0.0.255
一个子网	192.168.1.0/26	0.0.0.63
一个超网	192.168.0.0/16	0.0.255.255
所有 IP 地址	0.0.0.0/0	255.255.255.255

5.1.3 认识常用端口号

计算机之间依照 TCP/IP 进行通信，不同的协议对应不同的端口，例如，UDP（User Datagram Protocol，用户数据报协议）和 TCP（Transmission Control Protocol，传输控制协议）采用不同的端口号码。常用端口号如表 5-2 所示。

表 5-2 常用端口号

常见协议	常用端口号	协议基本作用
FTP	20 和 21	文件传输
SSH	22	安全远程登录
Telnet	23	远程登录
SMTP	25	邮件传输
DNS	53	域名解析
HTTP	80	超文本传输
POP3	110	邮件接收
HTTPS	443	加密超文本传输

5.1.4 了解 ACL 应用原则

由 5.1.1 节可知，ACL 主要作用是通过匹配报文实现对报文的过滤，从而完成对网络行为的管控。ACL 具有强大的功能，在使用时应该遵守如下原则。

- 每个接口、每个方向、每种协议只能应用一个 ACL（3P 原则）；
- 除命名 ACL 外，其余类型 ACL 无法单独删除某一条语句；
- ACL 语句末尾都隐含一条"拒绝所有"的语句，即 deny any；
- ACL 创建后需要应用于接口，否则不生效；
- ACL 用于过滤经过设备的报文，不会过滤设备本身产生的报文。

5.1.5 了解 ACL 应用范围

ACL 自问世以来，所应用的范围日趋广泛，许多协议因为 ACL 的加入变得更加安全、实用、灵活。现如今，ACL 已是企业网络中必不可少的安全工具，合理地应用 ACL 将会使网络结构更加稳定安全。表 5-3 展示了 ACL 的应用范围。

表 5-3 ACL 的应用范围

应用范围	应用技术
登录控制	Telnet、SSH、SNMP 等
报文过滤	QoS、NAT、IPSEC 等
路由策略	数据分流等
带宽控制	带宽预留等
提高安全性	拦截 DoS 攻击等
限制路由更新	分发列表

5.2 配置标准 ACL

5.2.1 认识标准 ACL

为了适应不同的场合，ACL 分为标准 ACL 和扩展 ACL 两类。其中，标准 ACL 根据数据包的源 IP 地址来决定允许或拒绝流量通过，其访问控制列表编号为 1～99 和 1300～1999。标准 ACL 的特点如表 5-4 所示。

表 5-4 标准 ACL 的特点

过滤依据	只检查数据包 IP 源地址
协议限制	整个 TCP/IP 协议集
编号范围	1～99、1300～1999
过滤效果	粗略，不能限制具体协议或端口
过滤速度	较快，检查参数少，仅检查源 IP 地址
应用位置	尽量靠近目标

5.2.2 学习标准 ACL 语法

标准 ACL 配置语法具体说明如下。

（1）依据编号定义标准 ACL

```
Router(config)#access-list access-list-number { deny | permit } source source-wildcard
```

命令解释如下：
- *access-list-number*：访问控制列表编号，编号范围为 1～99、1300～1999；
- **deny | permit**：拒绝／允许，决定符合条件的流量是否通过；
- *source*：数据包源地址，可以是主机地址，也可以是网络地址；
- *source-wildcard*：源地址通配符掩码，用来匹配源地址范围。

（2）依据名称定义标准 ACL

```
Router(config)#ip access-list standard { access-list-number | name }
Router(config-std-nacl)#{ deny | permit } source source-wildcard
```

命令解释如下：
- *standard*：标准 ACL，定义 ACL 的类型；
- *name*：访问控制列表名称，定义标准命名 ACL。

(3)应用标准 ACL

Router(config-if)#**ip access-group** { *access-list-number* | *name* } { **in** | **out** }

命令解释如下:
- **in** | **out**:入口 / 出口,定义 ACL 应用在接口上的方向。

(4)删除标准 ACL

Router(config)#**no access-list** *access-list-number*　　　　　//删除标准标号 ACL
Router(config)#**no ip access-list standard**{ *access-list-number* | *name* }　　//删除标准 ACL

5.2.3　场景一:配置标准编号 ACL

通过 5.2.1 节和 5.2.2 节内容的学习,相信大家已经对标准 ACL 有了初步认知,接下来让我们一起通过具体实验来完成标准编号 ACL 配置。标准编号 ACL 配置实验拓扑如图 5-1 所示。

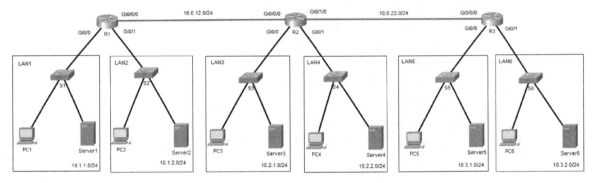

图 5-1　标准编号 ACL 配置实验拓扑

任务要求:
- 按图 5-1 所示搭建拓扑;
- 按图 5-1 所示为设备命名并配置 IP 地址;
- 配置 OSPF 路由使全网连通;
- 要求只允许 LAN1、LAN3 访问 LAN6,其余网段不做限制。

任务实施如下所述。

步骤一:为设备命名并配置接口 IP 地址

(1)在路由器 R1 上为设备命名并配置接口 IP 地址

```
Router(config)#hostname R1
R1(config)#interface GigabitEthernet0/0
R1(config-if)# ip address 10.1.1.1 255.255.255.0
```

```
R1(config-if)# no shutdown
R1(config-if)# interface GigabitEthernet0/1
R1(config-if)# ip address 10.1.2.1 255.255.255.0
R1(config-if)# no shutdown
R1(config-if)#interface GigabitEthernet0/0/0
R1(config-if)# ip address 10.0.12.1 255.255.255.0
R1(config-if)# no shutdown
```

（2）在路由器 R2 上为设备命名并配置接口 IP 地址

```
Router(config)#hostname R2
R2(config)#interface GigabitEthernet0/0
R2(config-if)# ip address 10.2.1.1 255.255.255.0
R2(config-if)# no shutdown
R2(config-if)#interface GigabitEthernet0/1
R2(config-if)# ip address 10.2.2.1 255.255.255.0
R2(config-if)# no shutdown
R2(config-if)#interface GigabitEthernet0/0/0
R2(config-if)# ip address 10.0.12.2 255.255.255.0
R2(config-if)# no shutdown
R2(config-if)#interface GigabitEthernet0/1/0
R2(config-if)# ip address 10.0.23.2 255.255.255.0
R2(config-if)# no shutdown
```

（3）在路由器 R3 上为设备命名并配置接口 IP 地址

```
Router(config)#hostname R3
R3(config)#interface GigabitEthernet0/0
R3(config-if)# ip address 10.3.1.1 255.255.255.0
R3(config-if)# no shutdown
R3(config-if)#interface GigabitEthernet0/1
R3(config-if)# ip address 10.3.2.1 255.255.255.0
R3(config-if)# no shutdown
R3(config-if)#interface GigabitEthernet0/0/0
R3(config-if)# ip address 10.0.23.3 255.255.255.0
R3(config-if)# no shutdown
```

步骤二：配置 OSPF 使全互通

（1）在路由器 R1 上配置 OSPF

> R1(config)#**router ospf 1**
> R1(config-router)#**router-id 1.1.1.1**
> R1(config-router)#**network 10.1.1.0 0.0.0.255 area 0**
> R1(config-router)#**network 10.1.2.0 0.0.0.255 area 0**
> R1(config-router)#**network 10.0.12.0 0.0.0.255 area 0**

（2）在路由器 R2 上配置 OSPF

> R2(config)#**router ospf 1**
> R2(config-router)#**router-id 2.2.2.2**
> R2(config-router)#**network 10.2.1.0 0.0.0.255 area 0**
> R2(config-router)#**network 10.2.2.0 0.0.0.255 area 0**
> R2(config-router)#**network 10.0.12.0 0.0.0.255 area 0**
> R2(config-router)#**network 10.0.23.0 0.0.0.255 area 0**

（3）在路由器 R3 上配置 OSPF

> R3(config)#**router ospf 1**
> R3(config-router)#**router-id 3.3.3.3**
> R3(config-router)#**network 10.3.1.0 0.0.0.255 area 0**
> R3(config-router)#**network 10.3.2.0 0.0.0.255 area 0**
> R3(config-router)#**network 10.0.23.0 0.0.0.255 area 0**

步骤三：查看路由表

在 R1 路由器上查看 OSPF 路由表：

> R1#**show ip route ospf**
>   10.0.0.0/8 is variably subnetted, 11 subnets, 2 masks
> O   10.0.23.0 [110/2] via 10.0.12.2, 00:09:06, GigabitEthernet0/0/0
> O   10.2.1.0 [110/2] via 10.0.12.2, 00:09:42, GigabitEthernet0/0/0
> O   10.2.2.0 [110/2] via 10.0.12.2, 00:01:17, GigabitEthernet0/0/0
> O   10.3.1.0 [110/3] via 10.0.12.2, 00:09:06, GigabitEthernet0/0/0
> O   10.3.2.0 [110/3] via 10.0.12.2, 00:01:34, GigabitEthernet0/0/0

步骤四：测试网络连通性

此时，LAN1～LAN5 均可以与 LAN6 通信。请依次进行连通性测试。

以下测试表明，PC1 可以成功访问 Server6。

> PC1:\>**ping 10.3.2.3**

Pinging 10.3.2.3 with 32 bytes of data:

Reply from 10.3.2.3: bytes=32 time=1ms TTL=125
Reply from 10.3.2.3: bytes=32 time<1ms TTL=125
Reply from 10.3.2.3: bytes=32 time=18ms TTL=125
Reply from 10.3.2.3: bytes=32 time<1ms TTL=125

Ping statistics for 10.3.2.3:
  Packets: Sent = 4, Received = 4, Lost = 0 (0% loss),
Approximate round trip times in milli-seconds:
  Minimum = 0ms, Maximum = 18ms, Average = 4ms

以下测试表明，PC4 可以成功访问 Server6。

PC4:\>**ping 10.3.2.3**

Pinging 10.3.2.3 with 32 bytes of data:

Reply from 10.3.2.3: bytes=32 time=1ms TTL=126
Reply from 10.3.2.3: bytes=32 time=11ms TTL=126
Reply from 10.3.2.3: bytes=32 time<1ms TTL=126
Reply from 10.3.2.3: bytes=32 time<1ms TTL=126

Ping statistics for 10.3.2.3:
  Packets: Sent = 4, Received = 4, Lost = 0 (0% loss),
Approximate round trip times in milli-seconds:
  Minimum = 0ms, Maximum = 11ms, Average = 3ms

步骤五：配置标准编号 ACL

在路由器 R3 上配置标准编号 ACL：

R3(config)#**access-list 10 permit 10.1.1.0 0.0.0.255**
R3(config)#**access-list 10 permit 10.2.1.0 0.0.0.255**
R3(config)#**access-list 10 deny any**
R3(config)#**interface GigabitEthernet0/1**
R3(config-if)#**ip access-group 10 out**

步骤六：查看标准编号 ACL

在路由器 R3 上查看标准编号 ACL：

```
R3#show access-lists
Standard IP access list 10
    10 permit 10.1.1.0 0.0.0.255
    20 permit 10.2.1.0 0.0.0.255
    30 deny any
```

步骤七：测试 ACL 应用效果

此时，LAN1 和 LAN3 中的主机可以访问 LAN6，但是 LAN2、LAN4 和 LAN5 中的主机被禁止访问 LAN6。请依次进行连通性测试。

以下测试表明，主机 PC1（LAN1）可以访问 Server6（LAN6）。

```
PC:\>ping 10.3.2.3

Pinging 10.3.2.3 with 32 bytes of data:

Reply from 10.3.2.3: bytes=32 time=1ms TTL=125
Reply from 10.3.2.3: bytes=32 time<1ms TTL=125
Reply from 10.3.2.3: bytes=32 time<1ms TTL=125
Reply from 10.3.2.3: bytes=32 time<1ms TTL=125

Ping statistics for 10.3.2.3:
    Packets: Sent = 4, Received = 4, Lost = 0 (0% loss),
Approximate round trip times in milli-seconds:
    Minimum = 0ms, Maximum = 1ms, Average = 0ms
```

以下测试表明，主机 PC2（LAN2）可以访问 Server6（LAN6）。

```
PC:\>ping 10.3.2.3

Pinging 10.3.2.3 with 32 bytes of data:

Reply from 10.0.23.3: Destination host unreachable.
Reply from 10.0.23.3: Destination host unreachable.
Reply from 10.0.23.3: Destination host unreachable.
Reply from 10.0.23.3: Destination host unreachable.

Ping statistics for 10.3.2.3:
    Packets: Sent = 4, Received = 0, Lost = 4 (100% loss),
```

5.2.4 场景二：配置标准命名 ACL

通过 5.2.3 节内容的学习，相信大家已经掌握了标准编号 ACL 配置，接下来，让我们一起通过具体实验来完成标准命名 ACL 配置。标准命名 ACL 配置实验拓扑如图 5-2 所示。

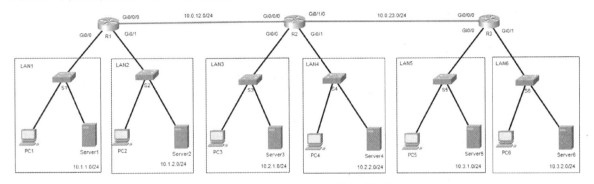

图 5-2 标准命名 ACL 配置实验拓扑

任务要求：
- 按图 5-2 所示搭建拓扑；
- 按图 5-2 所示为设备命名并配置 IP 地址；
- 配置 OSPF 路由使全网连通；
- 禁止 LAN1、LAN3 访问 LAN6，其余网段不做限制。

任务实施如下所述。

步骤一：为设备命名并配置接口 IP 地址

路由器 R1、R2 和 R3 的基础配置与 5.2.3 节场景一中的步骤一完全相同。

步骤二：配置 OSPF 使全网互通

路由器 R1、R2 和 R3 的 OSPF 配置与 5.2.3 节场景一中的步骤二完全相同。

步骤三：配置标准命名 ACL

在路由器 R3 上配置标准命名 ACL：

```
R3(config)#ip access-list standard NO-LAN6
R3(config-std-nacl)#deny 10.1.1.0 0.0.0.255
R3(config-std-nacl)#deny 10.2.1.0 0.0.0.255
R3(config-std-nacl)#permit any
R3(config-std-nacl)#interface GigabitEthernet0/1
R3(config-if)#ip access-group NO-LAN6 out
```

步骤四：查看标准命名 ACL

在路由器 R3 上查看标准命名 ACL：

```
R3#show access-lists
Standard IP access list NO-LAN6
    10 deny 10.1.1.0 0.0.0.255
    20 deny 10.2.1.0 0.0.0.255
    30 permit any
```

步骤五：测试 ACL 应用效果

此时，LAN1 和 LAN3 中的主机被禁止访问 LAN6，但 LAN2、LAN4 和 LAN5 中的主机可以访问 LAN6。请依次进行连通性测试。

以下测试表明，主机 PC1（LAN1）不能访问 Server6（LAN6）。

```
C:\>ping 10.3.2.3

Pinging 10.3.2.3 with 32 bytes of data:

Reply from 10.0.23.3: Destination host unreachable.
Reply from 10.0.23.3: Destination host unreachable.
Reply from 10.0.23.3: Destination host unreachable.
Reply from 10.0.23.3: Destination host unreachable.

Ping statistics for 10.3.2.3:
    Packets: Sent = 4, Received = 0, Lost = 4 (100% loss),
```

以下测试表明，主机 PC3（LAN3）也不能访问 Server6（LAN6）。

```
C:\>ping 10.3.2.3

Pinging 10.3.2.3 with 32 bytes of data:

Reply from 10.2.1.1: Destination host unreachable.
Reply from 10.2.1.1: Destination host unreachable.
Reply from 10.2.1.1: Destination host unreachable.
Reply from 10.2.1.1: Destination host unreachable.

Ping statistics for 10.3.2.3:
    Packets: Sent = 4, Received = 0, Lost = 4 (100% loss),
```

5.3 配置扩展 ACL

5.3.1 认识扩展 ACL

前面介绍了标准 ACL 相关知识，接下来介绍扩展 ACL 相关内容。与标准 ACL 不同的是，扩展 ACL 根据数据包的源/目的 IP 地址、协议和端口来决定允许或拒绝流量通过。扩展 ACL 的访问控制列表号为 100～199、2000～2699。从我们对标准 ACL 和扩展 ACL 的描述中不难发现，两者各有特点，表 5-5 展示了扩展 ACL 的特点。

表 5-5 扩展 ACL 的特点

过滤依据	同时检查源 IP 地址和目的 IP 地址
协议限制	限制具体协议或端口号
编号范围	100～199，2000～2699
过滤能力	强，可覆盖标准 ACL 的功能
过滤速度	慢，匹配 IP 地址、协议或端口
应用位置	尽量靠近被拒绝的源

为了灵活应用扩展 ACL，人们为扩展 ACL 定义了运算符，以便其更好地匹配流量。ACL 运算符及其描述如表 5-6 所示。

表 5-6 ACL 运算符及其描述

ACL 运算符	运算符描述
lt	小于（less than）
gt	大于（greater than）
eq	等于（equal to）
neq	不等于（not equal to）
range	端口范围

5.3.2 学习扩展 ACL 语法

扩展 ACL 配置语法具体说明如下。

（1）依据编号定义扩展 ACL

Router(config)#**access-list** *access-list-number* { **deny** | **permit** } *protocol source source-wildcard* [*operator port*] *destination destination-wildcard* [*operator port*]

命令解释如下：
- *protocol*：协议，用来指定协议类型，例如，IP、TCP 和 UDP 等；

- *operator*：ACL 运算符，例如，lt、gt 和 eq 等；
- *port*：端口编号，编号范围为 0～65535 或采用 ftp、telnet、www 等；
- *destination*：数据包目的地址，可以是主机地址，也可以是网络地址；
- *destination-wildcard*：目的地址通配符掩码，用来匹配目的地址范围。

（2）依据名称定义扩展 ACL

Router(config)#**ip access-list extended** { *access-list-number* | *name* }
Router(config-ext-nacl)#{ **deny** | **permit** } *protocol source source-wildcard* [*operator port*] *destination destination-wildcard* [*operator port*]

（3）应用扩展 ACL

Router(config-if)#**ip access-group** { *access-list-number* | *name* } {**in**|**out**}

（4）删除扩展 ACL

Router(config)#**no access-list** *access-list-number*　　　　　　　　　　//删除扩展标号 ACL
Router(config)#**no ip access-list extended** { *access-list-number* | *name* }　　//删除扩展 ACL

5.3.3　场景三：配置扩展编号 ACL

通过 5.3.1 节和 5.3.2 节内容的学习，相信大家都已经对扩展 ACL 有了初步认知。接下来，让我们一起通过具体实验来完成扩展编号 ACL 配置。扩展编号 ACL 实验拓扑如图 5-3 所示。

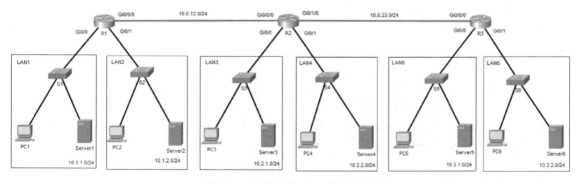

图 5-3　扩展编号 ACL 实验拓扑

任务要求：

- 按图 5-3 所示搭建拓扑；
- 按图 5-3 所示为设备命名并配置 IP 地址；
- 配置 OSPF 路由使全网连通；
- 禁止 LAN1 和 LAN3 中的主机发给 Server6 的 ping 包，其余不做限制。

任务实施如下所述。

步骤一：为设备命名并配置接口 IP 地址

路由器 R1、R2 和 R3 的基础配置与 5.2.3 节场景一中的步骤一完全相同。

步骤二：配置 OSPF 使全网互通

路由器 R1、R2 和 R3 的 OSPF 配置与 5.2.3 节场景一中的步骤二完全相同。

步骤三：配置扩展编号 ACL

（1）在路由器 R1 上配置扩展编号 ACL

```
R1(config)#access-list 110 deny icmp 10.1.1.0 0.0.0.255 10.3.2.3 0.0.0.0
R1(config)#access-list 110 permit ip any any
R1(config)#interface GigabitEthernet0/0
R1(config-if)#ip access-group 110 in
```

（2）在路由器 R2 上配置扩展编号 ACL

```
R2(config)#access-list 110 deny icmp 10.2.1.0 0.0.0.255 10.3.2.3 0.0.0.0
R2(config)#access-list 110 permit ip any any
R2(config)#interface GigabitEthernet0/0
R2(config-if)#ip access-group 110 in
```

步骤四：查看 ACL

（1）在路由器 R1 上查看 ACL

```
R1#show access-lists
Extended IP access list 110
    10 deny icmp 10.1.1.0 0.0.0.255 host 10.3.2.3
    20 permit ip any any
```

（2）在路由器 R2 上查看 ACL

```
R2#show access-lists
Extended IP access list 110
    10 deny icmp 10.2.1.0 0.0.0.255 host 10.3.2.3
    20 permit ip any any
```

步骤五：测试 ACL 应用效果

此时，LAN1 和 LAN3 中的主机发给 Server6 的 ping 包将被拦截，但是，Server6 不会拦截 LAN2、LAN4、LAN5 和 LAN6 中的主机发送的 ping 包。请依次进行连通性测试。

以下测试表明，主机 PC1（LAN1）可以 ping 通主机 PC6（LAN6）。

```
C:\>ping 10.3.2.2

Pinging 10.3.2.2 with 32 bytes of data:

Reply from 10.3.2.2: bytes=32 time=1ms TTL=125
Reply from 10.3.2.2: bytes=32 time<1ms TTL=125
Reply from 10.3.2.2: bytes=32 time<1ms TTL=125
Reply from 10.3.2.2: bytes=32 time=1ms TTL=125

Ping statistics for 10.3.2.2:
    Packets: Sent = 4, Received = 4, Lost = 0 (0% loss),
Approximate round trip times in milli-seconds:
    Minimum = 0ms, Maximum = 1ms, Average = 0ms
```

以下测试表明，主机 PC1（LAN1）不能 ping 通 Server6（LAN6）。

```
C:\>ping 10.3.2.3

Pinging 10.3.2.3 with 32 bytes of data:

Reply from 10.1.1.1: Destination host unreachable.
Reply from 10.1.1.1: Destination host unreachable.
Reply from 10.1.1.1: Destination host unreachable.
Reply from 10.1.1.1: Destination host unreachable.

Ping statistics for 10.3.2.3:
    Packets: Sent = 4, Received = 0, Lost = 4 (100% loss),
```

扩展 ACL 的配置非常灵活，因为它需要同时基于源地址和目的地址匹配数据包，所以扩展 ACL 的创建可以根据实际应用场景灵活进行。在本场景中，分别在路由器 R1 和 R2 上配置了 ACL 并达到了预期效果。这是不是最佳方案，取决于在实际应用环境中你对路由器的控制权限。如果你只能控制路由器 R3，请根据实际情况选择最佳方案。

5.3.4　场景四：配置扩展命名 ACL

通过 5.3.3 节内容的学习，相信大家已经掌握扩展编号 ACL 的配置，接下来，让我们一起通过具体实验来完成扩展命名 ACL 配置。扩展命名 ACL 实验拓扑如 5-4 所示。

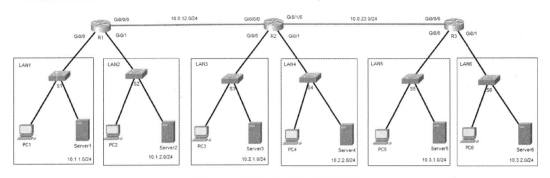

图 5-4 扩展命名 ACL 实验拓扑

任务要求：
- 按图 5-4 所示搭建拓扑；
- 按图 5-4 所示为设备命名并配置 IP 地址；
- 配置 OSPF 路由使全网连通；
- 禁止 LAN1、LAN3 访问 Server6 的 WEB 页面，其余不做限制。

任务实施如下所述。

步骤一：为设备命名并配置接口 IP 地址

路由器 R1、R2 和 R3 的基础配置与 5.2.3 节场景一中的步骤一完全相同。

步骤二：配置 OSPF 使全网互通

路由器 R1、R2 和 R3 的 OSPF 配置与 5.2.3 节场景一中的步骤二完全相同。

步骤三：配置扩展命名 ACL

（1）在路由器 R1 上配置扩展命名 ACL

```
R1(config)#ip access-list extended NO-Server6
R1(config-ext-nacl)#deny tcp 10.1.1.0 0.0.0.255 10.3.2.3 0.0.0.0 eq www
R1(config-ext-nacl)#permit ip any any
R1(config-ext-nacl)#interface GigabitEthernet0/0
R1(config-if)#ip access-group NO-Server6 in
```

（2）在路由器 R2 上配置扩展命名 ACL

```
R2(config)#ip access-list extended NO-Server6
R2(config-ext-nacl)#deny tcp 10.2.1.0 0.0.0.255 10.3.2.3 0.0.0.0 eq www
R2(config-ext-nacl)#permit ip any any
R2(config-ext-nacl)#interface GigabitEthernet0/0
```

R2(config-if)#**ip access-group NO-Server6 in**

步骤四：查看 ACL

（1）在路由器 R1 上查看 ACL

R1#**show access-lists**
Extended IP access list NO-Server6
　　10 deny tcp 10.1.1.0 0.0.0.255 host 10.3.2.3 eq www
　　20 permit ip any any

（2）在路由器 R2 上查看 ACL

R2#**show access-lists**
Extended IP access list NO-Server6
　　10 deny tcp 10.2.1.0 0.0.0.255 host 10.3.2.3 eq www
　　20 permit ip any any

步骤五：测试 ACL 应用效果

此时，LAN1 和 LAN3 中的主机不能打开 Server6 的 WEB 页面，但可以 ping 通 Server6；LAN2、LAN4、LAN5 和 LAN6 中的主机可以正常访问 Server6。请使用浏览器和 **ping** 命令按要求依次进行访问测试。PC1 和 PC2 访问 Server6 的测试结果分别如图 5-5 和图 5-6 所示，图 5-5 展示了主机 PC1（LAN1）不能打开 Server6 的 Web 页面，图 5-6 展示了主机 PC2（LAN2）能正常打开 Server6 的 WEB 页面。

图 5-5　PC1 访问 Server6 的测试结果

图 5-6　PC2 访问 Server6 的测试结果

请思考，假设你是网络工程师，你有对所有设备的控制权限，以上扩展命名 ACL 实施方案是否为最佳实践方案？如果不是，请说明如何修改，给出你的设计实施方案。

5.4 拓展 ACL 应用

5.4.1 认识 VTY 技术

VTY（Virtual Type Terminal，虚拟类型终端）是 Cisco 设备管理的一种方式，用来定义同时远程登录的设备数量。例如，vty 0 4 表示拥有 0、1、2、3、4 五个端口，可以同时承载 5 台远程设备同时登录。

5.4.2 场景五：采用 ACL 限制 VTY 访问

通过本章内容的学习，我们发现 ACL 的应用非常广泛。接下来，让我们一起来了解 ACL 在远程登录方面的应用。ACL 限制 VTY 访问实验拓扑如图 5-7 所示。

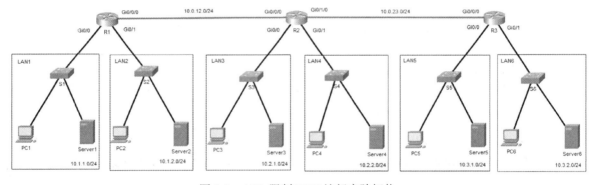

图 5-7　ACL 限制 VTY 访问实验拓扑

任务要求：
- 按图 5-7 所示搭建拓扑；
- 按图 5-7 所示为设备命名并配置 IP 地址；
- 配置 OSPF 路由使全网连通；
- 只允许 LAN1 和 LAN2 中的主机 Telnet 路由器 R3。

任务实施如下所述。

步骤一：为设备命名并配置接口 IP 地址

路由器 R1、R2 和 R3 的基础配置与 5.2.3 节场景一中的步骤一完全相同。

步骤二：配置 OSPF 使全网互通

路由器 R1、R2 和 R3 的 OSPF 配置与 5.2.3 节场景一中的步骤二完全相同。

步骤三：配置 VTY

在路由器 R3 上配置 VTY：

```
R3(config)#line vty 0 4
R3(config-line)#password cisco
R3(config-line)#login
```

步骤四：配置 ACL

在路由器 R3 上配置 ACL：

```
R3(config)#access-list 1 permit 10.1.1.0 0.0.0.255
R3(config)#access-list 1 permit 10.1.2.0 0.0.0.255
R3(config)#access-list 1 deny any
```

步骤五：在 VTY 线路上应用 ACL

在路由器 R3 上配置 VTY：

```
R3(config)#line vty 0 4
R3(config-line)#access-class 1 in
```

步骤六：查看 ACL 和 VTY

（1）在路由器 R3 上查看 ACL

```
R3#show access-lists
Standard IP access list 1
    10 permit 10.1.1.0 0.0.0.255
    20 permit 10.1.2.0 0.0.0.255
    30 deny any
```

（2）在路由器 R3 上查看 VTY

```
R3#show line
   Tty Line Typ      Tx/Rx     A  Roty AccO AccI Uses  Noise Overruns  Int
 *   0    0 CTY                -   -    -    -     0     0    0/0       -
     1    1 AUX   9600/9600    -   -    -    -     0     0    0/0       -
 * 388  388 VTY                -   -    -    1     0     0    0/0       -
```

* 389	389 VTY	-	-	1	0	0	0/0	-
390	390 VTY	-	-	1	0	0	0/0	-
391	391 VTY	-	-	1	0	0	0/0	-
392	392 VTY	-	-	1	0	0	0/0	-

Line(s) not in async mode -or- with no hardware support:
3-387

步骤七：测试 ACL 应用效果

此时，LAN1 和 LAN2 中的主机均可以通过 Telnet 远程登录路由器 R3，而 LAN3～LAN6 中的主机被限制登录，请依次进行登录测试。

以下测试表明，主机 PC1（LAN1）成功登录路由器 R3：

C:\>**telnet 10.3.2.1**

Trying 10.3.2.1 ...Open

User Access Verification

Password:
R3>

以下测试表明，主机 PC3（LAN3）被路由器 R3 拒绝登录：

C:\>**telnet 10.3.1.1**

Trying 10.3.1.1 ...
% Connection refused by remote host

5.4.3 认识 QoS 技术

QoS（Quality of Service，服务质量）是一种应用非常广泛的安全、控制机制，其目的是为网络通信提供更好的服务能力，针对不同的数据流采用不同的优先级，优化数据流的性能，为其提供端到端的服务质量保证。QoS 主要应用于解决网络延迟和阻塞问题，在应用 QoS 时，可以先通过 ACL 匹配数据流，然后再对其定义规则与策略。

5.4.4 场景六：将 ACL 应用于 QoS

在简单了解 QoS 原理之后，接下来，我们将一起通过具体实验来进一步了解 ACL 在 QoS 中的应用。ACL 应用于 QoS 实验拓扑如图 5-8 所示。

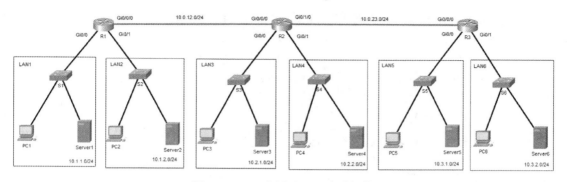

图 5-8 ACL 应用于 QoS 实验拓扑

任务要求：
- 按图 5-8 所示搭建拓扑；
- 按图 5-8 所示为设备命名并配置 IP 地址；
- 配置 OSPF 路由使全网连通；
- 通过配置 QoS 限制 LAN1 的带宽。

任务实施如下所述。

步骤一：为设备命名并配置接口 IP 地址

路由器 R1、R2 和 R3 的基础配置与 5.2.3 节场景一中的步骤一完全相同。

步骤二：配置 OSPF 使全网互通

路由器 R1、R2 和 R3 的 OSPF 配置与 5.2.3 节场景一中的步骤二完全相同。

步骤三：配置 QoS

在路由器 R1 上配置 QoS：

```
R1(config)#access-list 110 permit ip 10.1.1.0 0.0.0.255 any
R1(config)#class-map match-any 110               //创建满足任意条件的类映射表
R1(config-cmap)#class-map match-any class-1
R1(config-cmap)# match access-group 110          //匹配 ACL
R1(config-cmap)#policy-map policy-1              //创建规则映射表
R1(config-pmap)# class class-1                   //引用类映射表
R1(config-pmap-c)#priority percent 75            //指定带宽为接口带宽的 75%
R1(config-pmap-c)#interface Gig0/0/0
R1(config-if)# service-policy output policy-1    //应用到接口
```

步骤四：查看映射表信息

（1）在路由器 R1 上查看类映射表信息

```
R1#show class-map
```

```
Class Map match-any class-default (id 0)
  Match any
Class Map match-any 110 (id 1)
  Match none
Class Map match-any class-1 (id 2)
  Match access-group 110
```

（2）在路由器 R1 上查看规则映射表信息

```
R1#show policy-map
  Policy Map policy-1
    Class class-1
      Strict Priority
      Bandwidth 75 (%)
```

5.5 挑战练习

5.5.1 挑战练习一

挑战要求：通过本章的学习，我们发现 ACL 应用非常广泛，它还有许多未知应用等待我们去探索。例如，ACL 在路由更新方面的应用。接下来，让我们一起挑战 17NET2 班 XL 同学设计的击剑实验，该实验拓扑如图 5-9 所示。

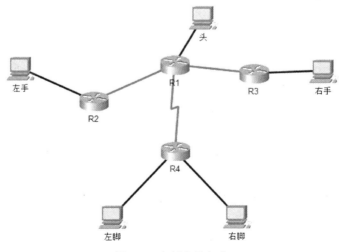

图 5-9　击剑实验拓扑

任务要求：
- 按图 5-9 所示搭建拓扑；
- 按图 5-9 所示为设备命名并配置 IP 地址；
- 配置 OSPF 路由使全网连通；
- 配置扩展 ACL 禁止 OSPF 路由更新；
- 查看路由表，发现 OSPF 学到的路由消失。

5.5.2 挑战练习二

挑战要求：通过前面 5.5.1 节的挑战练习，我们发现 ACL 可以应用在路由更新中，但其应用远不止于此。我们可以利用 ACL 通配符掩码特性（0 为匹配，1 为忽略）来设置灵活的策略，例如，ACL 在"奇偶性"方面的扩展，ACL 奇偶性匹配如表 5-7 所示。接下来，让我们一起挑战如下的坦克实验，该实验拓扑如图 5-10 所示。

表 5-7　ACL 奇偶性匹配

匹配原则	ACL 语句	匹配地址 / 通配符掩码
匹配奇数	access-list 1 permit 192.168.5.0 0.0.254.0	11000000.10101000.00000101.00000000（192.168.5.0） 00000000.00000000.11111110.00000000（0.0.254.0）
匹配偶数	access-list 1 permit 192.168.6.0 0.0.254.0	11000000.10101000.00000110.00000000（192.168.6.0） 00000000.00000000.11111110.00000000（0.0.254.0）

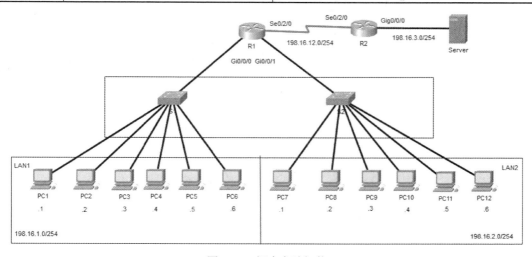

图 5-10　坦克实验拓扑

任务要求：
- 按图 5-10 所示搭建拓扑；
- 按图 5-10 所示为设备命名并配置 IP 地址；

- 配置 OSPF 路由使全网连通；
- 配置 ACL，Server 只允许 LAN1 中奇数号主机和 LAN2 中偶数号主机以 WEB 方式访问；
- 配置 ACL，使 R1 仅允许 LAN1 中的主机通过 SSH 远程登录。

5.6 本章小结

本章内容到此结束。本章主要内容包括 ACL 概述、标准 ACL 和扩展 ACL 以及 ACL 拓展应用。在整个网络体系中 ACL 起到了非常重要的作用，可以用"无处不在"来形容，也可以用网络中的"润滑剂"来形容，例如，限制 VTY 访问和在 QoS 中的应用。ACL 是提高网络安全性的一个有效工具，它的应用使得网络更加健壮，在第 6 章中，我们将使用 ACL 来限制私有网络对公有网络的访问。本章设计了 6 个应用场景和 2 个挑战练习，目的是让读者能够熟练应用 ACL 技术，提高网络的安全性。ACL 的应用还有很多，有兴趣的读者可以自行了解。

第 6 章

学习 NAT 技术

本章要点：

- 认识 NAT 技术
- 配置多种类型的 NAT
- 配置拓展 NAT
- 挑战练习
- 本章小结

本章通过介绍网络地址转换（NAT）技术，使读者能够理解 NAT 的作用——有效缓解 IPv4 地址短缺状况，实现内部网络与外部网络的通信。其中，6.1 节介绍 NAT 的定义、分类、术语、原理以及优缺点；6.2 节介绍多种类型 NAT 的配置，包括动态 NAT、动态 PAT、静态 NAT 以及静态 PAT 的配置；6.3 节介绍拓展 NAT，包括无线路由器内置 NAT 的配置以及由外至内映射的 NAT 配置。本章设计了 6 个应用场景和 2 个挑战练习，使读者能够熟练应用 NAT 技术，实现私有网络（简称私网）对公共网络（简称公网）的访问。

6.1 认识 NAT 技术

6.1.1 理解 NAT 的定义

NAT（Network Address Translation，网络地址转换）技术可将 IP 数据报文头中的 IP 地址转换为另一个 IP 地址，其主要作用是将内部私有网络地址转换为外部公共网络地址，适用于解决 IP 地址短缺问题。

6.1.2 熟悉 NAT 的分类

NAT 有 3 种类型：静态 NAT、动态 NAT 和 NAT 过载（PAT）。
① 静态 NAT：将内部本地地址与内部全局地址一对一地映射，实现静态地址转换。
② 动态 NAT：将内部本地地址与内部全局地址一对多地映射，实现动态地址转换。动态 NAT 首先要定义合法 IP 地址池，然后采用动态分配的方式将合法 IP 地址池中的地址一对一地映射为内部 IP 地址。
③ PAT（端口地址转换）：将内部本地地址与内部全局地址多对一地映射，实现端口地址转换。PAT 又称 NAT 过载，可把内部地址映射到外部网络 IP 地址的不同端口上，从而可以实现多对一的映射。

6.1.3 了解 NAT 的术语

在采用 NAT 技术进行通信时，可根据网络地址类型、流量传输方向，使用不同的 IPv4 地址。在 NAT 模式下，可用以下 4 种称谓来描述这些地址。
● 内部本地地址：转换前的内部源地址；
● 内部全局地址：转换后的内部主机地址；
● 外部本地地址：转换前的外部目标主机地址；
● 外部全局地址：转换后的外部目标主机地址。
为了更好地理解 NAT 的术语，需要首先理解内部地址与外部地址、本地地址与全局地址的基本概念。
- 内部地址：经过 NAT 转换的设备地址；
- 外部地址：目的设备的地址；
- 本地地址：在网络内部出现的地址；
- 全局地址：在网络外部出现的地址。

所谓不同的 NAT 称谓不过是这些基本术语的组合。下面我们通过一个例子来进一步了解这些 NAT 称谓。NAT 术语说明拓扑如图 6-1 所示，具体说明如下。

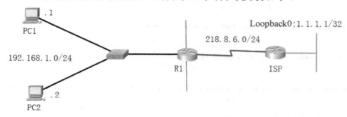

图 6-1　NAT 术语说明拓扑

① 内部本地地址是在网络内部看到的源地址。如图 6-1 所示，192.168.1.1 与 192.168.1.2 分别为 PC1 与 PC2 的内部本地地址。

② 内部全局地址是在外部网络看到的源地址，换言之，是内部本地地址采用 NAT 技术转换后的地址。如图 6-1 所示，在路由器 R1 处，将源地址 192.168.1.1 转换成 218.8.6.1，用 NAT 的术语来说就是，在路由器 R1 处，将内部本地地址 192.168.1.1 转换成为内部全局地址 218.8.6.1。

③ 外部全局地址是从外部网络看到的目的地址，它是 Internet 上主机的 IP 地址。如图 6-1 所示，环回接口地址 1.1.1.1/32 即为外部全局地址。

④ 外部本地地址是从外部网络内部看到的目的地址。如图 6-1 所示，PC1 将流量发送到 IPv4 地址为 1.1.1.1/32 的环回接口，1.1.1.1/32 也就是外部本地地址。

6.1.4　理解 NAT 的原理

NAT 运行原理如图 6-2 所示，在图中，内网地址为 192.168.10.1 的 PC1 想要访问具有公网地址 217.12.16.1 的 WEB 服务器，借助该图，我们来学习一下 NAT 的运行原理。

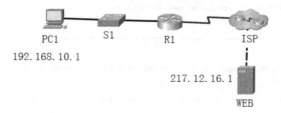

图 6-2　NAT 运行原理

PC1 发送了一个目的地为 WEB 服务器的数据包,该数据包经路由器 R1 进入 ISP,如图 6-3 所示。

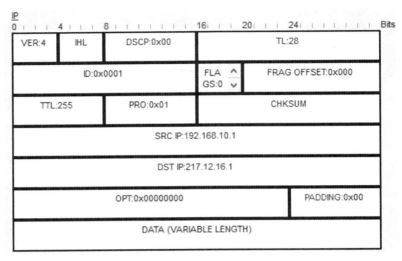

图 6-3 数据包经路由器 R1 进入 ISP

当数据包到达路由器 R1（该路由器启用了 NAT 功能）时,路由器 R1 会读取数据包的源 IPv4 地址,然后将 192.168.10.1（内部本地地址）转换为 218.8.7.10（内部全局地址）。路由器 R1 将此本地与全局地址的映射关系添加到 NAT 表中。

路由器 R1 将具有转换后的源地址的数据包发送到目的地,如图 6-4 所示,地址被转换为内部全局地址。

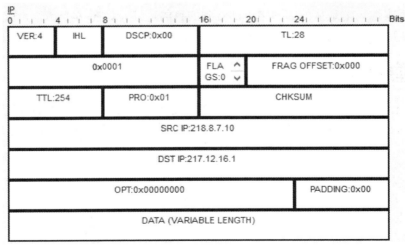

图 6-4 地址被转换为内部全局地址

WEB 服务器以一个目的地址为 PC1 的内部全局地址（218.8.7.10）的数据包做出响应。响应数据包如图 6-5 所示。

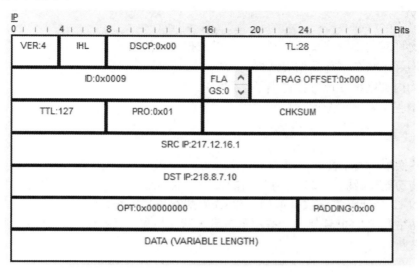

图 6-5　响应数据包

路由器 R1 收到这个目的地址为 218.8.7.10 的数据包后将检查 NAT 表，找出有关此映射的条目，路由器 R1 使用此信息将内部全局地址（218.8.7.10）转换为内部本地地址（192.168.10.1），然后将数据包转发到 PC1，转发数据包如图 6-6 所示。

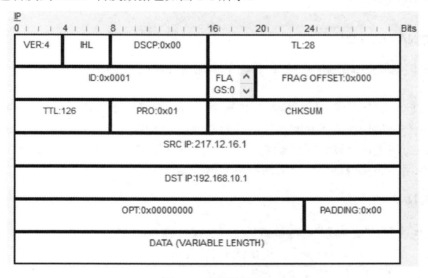

图 6-6　转发数据包

6.1.5 理解 NAT 的优缺点

1. NAT 优点

- NAT 允许对内部网络实行私网寻址，以维护合法注册的公网寻址方案；
- NAT 增强了与公网连接的灵活性；
- NAT 为内部网络寻址方案提供了一致性；
- NAT 增强了网络安全性。

2. NAT 缺点

- 地址转换影响网络性能，增加了交换延迟；
- 端到端功能被减弱，端到端寻址功能丧失，端到端 IPv4 可追溯性也会丧失；
- 采用 NAT 会使隧道协议（例如，IPSec）配置更加复杂；
- 外部网络发起 TCP 连接的一些服务或者无状态协议（诸如使用 UDP 的无状态协议）可能会中断。

6.2 配置多种类型的 NAT

6.2.1 认识动态 NAT

1. 动态 NAT 的定义

动态 NAT（动态网络地址转换）：将内部本地地址与内部全局地址实现一对一的不固定的动态映射。

动态 NAT 将多个公网地址放置在地址池中，当私网主机要访问公网时，路由器会从地址池中提取一个公网地址作为该私网地址的映射地址。当已使用的公网地址在一定时间内没有数据传输时，该公网地址会被收回到地址池中，以供下次使用。动态 NAT 并不会节省公网地址，所以其主要应用于隐藏内部地址以实现网络安全的场合。

2. 动态 NAT 的语法

Router(config)#**ip nat pool** *name start-ip end-ip* **netmask** *netmask*

- ***name***：名称，定义 nat pool（地址池）名称；
- ***start-ip***：开始地址，定义地址池开始地址；
- ***end-ip***：结束地址，定义地址池结束地址；

- *netmask*：子网掩码，定义地址池中的 IP 子网掩码。

> Router(config)#**access-list** *access-list-number* **permit** *source [source-wildcard]*

配置标准访问列表，允许应转换的地址：

> Router(config)#**ip nat inside source list** *access-list-number* **pool** *name*

通过绑定已定义的访问列表和地址池，实现动态源地址转换：

> Router(config-if)#**ip nat { inside | outside}**

- **inside | outside**：内部或外部接口，定义 nat 的应用方向。

当没有数据传输时，该公网地址会被收回到地址池中，以供下次使用。

6.2.2 场景一：配置动态 NAT

某公司使用 172.16.10.0/24 网段为员工分配 IP 地址，现公司员工需要访问外网，已知该公司拥有 218.8.1.1/24、218.8.1.2/24、218.8.1.3/24、218.8.1.4/24 四个公网地址，网络工程师希望通过配置公网地址池，使员工采用动态 NAT 方式访问外网，配置动态 NAT 实验拓扑如图 6-7 所示。

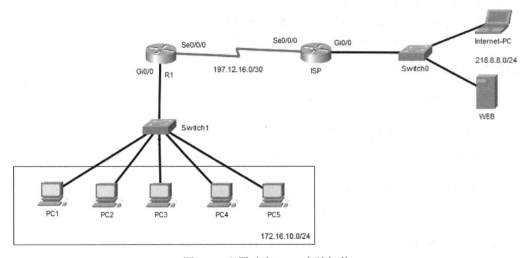

图 6-7　配置动态 NAT 实验拓扑

任务要求：

- 按图 6-7 所示搭建拓扑；
- 按图 6-7 所示为设备命名并配置 IP 地址；
- 配置合适的路由使得全网连通；
- 配置动态 NAT 使得内网可以访问公网。

任务实施如下所述。

步骤一：为设备命名并配置接口 IP 地址

（1）在路由器 R1 上为设备命名并配置接口 IP 地址

```
Router(config)#hostname R1
R1(config)#interface GigabitEthernet0/0
R1(config-if)#ip address 172.16.10.254 255.255.255.0
R1(config-if)#no shutdown
R1(config-if)#interface Serial0/0/0
R1(config-if)#ip address 197.12.16.1 255.255.255.252
R1(config-if)#no shutdown
```

（2）在路由器 ISP 上为设备命名并配置接口 IP 地址

```
Router(config)#hostname ISP
ISP(config)#interface GigabitEthernet0/0
ISP(config-if)#ip address 218.8.8.254 255.255.255.0
ISP(config-if)#no shutdown
ISP(config-if)#interface Serial0/0/0
ISP(config-if)#ip address 197.12.16.2 255.255.255.252
ISP(config-if)#no shutdown
```

步骤二：配置动态 NAT

（1）在路由器 R1 上配置静态默认路由

```
R1(config)# ip route 0.0.0.0 0.0.0.0 197.12.16.2
```

（2）在路由器 R1 上创建 ACL 并标识能够进行转换的地址

```
R1(config)# access-list 1 permit 172.16.10.0 0.0.0.255
```

（3）在路由器 R1 上配置 NAT 地址池

```
R1(config)# ip nat pool NAT-POOL 218.8.1.1 218.8.1.4 netmask 255.255.255.0
```

（4）在路由器 R1 上绑定 ACL 与地址池

```
R1(config)# ip nat inside source list 1 pool NAT-POOL
```

（5）在路由器 R1 上配置 NAT 的内部接口与外部接口

```
R1(config)# interface GigabitEthernet 0/0
R1(config-if)#ip nat inside
```

```
R1(config)#interface serial 0/0/0
R1(config-if)#ip nat outside
```

（6）在路由器 ISP 上配置到达对端公网地址的静态路由

```
ISP(config)#ip route 218.8.1.0 255.255.255.0 197.12.16.1
```

步骤三：验证动态 NAT 映射

检验动态 NAT：

```
R1#show ip nat translations
Pro   Inside global      Inside local      Outside local      Outside global
icmp  218.8.1.1:2        172.16.10.2:2     218.8.8.100:2      218.8.8.100:2
icmp  218.8.1.2:2        172.16.10.3:2     218.8.8.100:2      218.8.8.100:2
icmp  218.8.1.3:2        172.16.10.4:2     218.8.8.100:2      218.8.8.100:2
icmp  218.8.1.4:8        172.16.10.1:8     218.8.8.100:8      218.8.8.100:8
```

由上面 **show ip nat translations** 命令的输出结果可知，所有已经配置的静态转换和为所有流量创建的动态转换均已实现。

```
R1#show ip nat statistics
Total translations: 4 (0 static, 4 dynamic, 4 extended)
Outside Interfaces: Serial0/0/0
Inside Interfaces: GigabitEthernet0/0
Hits: 17    Misses: 531
Expired translations: 14
Dynamic mappings:
-- Inside Source
access-list 1 pool NAT-POOL refCount 4
 pool NAT-POOL: netmask 255.255.255.0
        start 218.8.1.1 end 218.8.1.4
        type generic, total addresses 4 , allocated 4 (100%), misses 1
```

show ip nat statistics 命令可用于显示有关动态地址转换条目、NAT 配置参数、地址池中地址数量和已经分配地址数量等信息。

由上面 **show ip nat statistics** 命令输出结果可知，当前有 4 个动态 NAT 进程正在进行，当前地址池中共有 4 个地址且已经全被占用。

步骤四：网络连通性测试

测试结果如图 6-8 所示，由该图可知，测试顺利通过，内网中除了 PC5，所有 PC 均可以访问外网中的 WEB 服务器和 PC。

Fire	Last Status	Source	Destination	Type	Color	Time(sec)	Periodic	Num	Edit	Delete
●	Successful	PC1	WEB	ICMP		0.000	N	0	(edit)	(delete)
●	Successful	PC2	WEB	ICMP		0.000	N	1	(edit)	(delete)
●	Successful	PC3	WEB	ICMP		0.000	N	2	(edit)	(delete)
●	Successful	PC4	WEB	ICMP		0.000	N	3	(edit)	(delete)
●	Failed	PC5	WEB	ICMP		0.000	N	4	(edit)	(delete)

图 6-8　测试结果

那么问题来了，为什么整个网络唯有 PC5 不能成功访问外网 WEB 服务器？我们通过 **show ip nat statistics** 命令查看地址池，发现地址池中所有地址均被分配出去，也就是说 PC5 无法从地址池中获取转换后的公网 IP 地址。我们可以通过 **clear ip nat translation *** 命令清除 NAT 表中所有动态地址转换条目，然后再尝试一下 PC5 是否可以连接外网，如图 6-9 所示，PC5 与外网 WEB 服务器连接成功。

Fire	Last Status	Source	Destination	Type	Color	Time(sec)	Periodic	Num	Edit	Delete
●	Successful	PC5	WEB	ICMP		0.000	N	0	(edit)	(delete)

图 6-9　PC5 与外网 WEB 服务器连接成功

由图 6-9 可知，PC5 可以与外网 WEB 服务器正常通信，并由此得出如下结论：
- 地址池中的地址遵循先到先得的原则；
- 地址池中 IP 地址的数量决定了可以与外网通信的 PC 的数量。

6.2.3　认识动态 PAT

1．动态 PAT 的定义

动态 PAT（动态端口地址转换）：将内部本地地址与内部全局地址实现多对一不固定映射。

动态 PAT 可以使多个内部本地地址共享一个公网 IP 地址，通过 IP 地址+端口方式识别不同的源地址。PAT 适用于公网地址远少于私网地址的网络。由于这种地址转换方式对公网地址需求少，因此应用最为广泛，其特点是私网主机可以访问公网，反之则不可以。

2．动态 PAT 的语法

（1）基于地址池的 PAT

```
Router(config)#ip nat pool name start-ip end-ip netmask netmask
Router(config)#access-list access-list-number permit source [source-wildcard]
Router(config)#ip nat inside source list access-list-number pool name overload
```

- overload：过载转换，将定义的访问控制列表与地址池建立过载转换。

```
Router(config-if)#ip nat { inside |outside}
```

（2）基于端口的 PAT

```
Router(config)#access-list access-list-number permit source [source-wildcard]
Router(config)#ip nat inside source list access-list-number interface interface overload
Router(config-if)#ip nat { inside|outside }
```

6.2.4 场景二：配置动态 PAT

通过 6.2.3 节内容的学习，我们知道动态 NAT 受到公网地址池中 IP 地址数量的限制，但是目前公网 IPv4 地址数量紧缺，配置公网地址池较浪费公网地址。PAT 又称 NAT 过载，可将内部地址映射到外部网络 IP 地址的不同端口上，从而可以实现多对一的映射，是节省 IPv4 地址最为有效的方法。NAT 过载实验拓扑如图 6-10 所示。

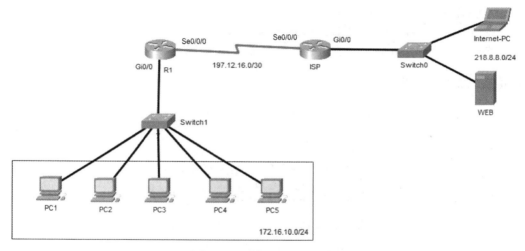

图 6-10　NAT 过载实验拓扑

任务要求：
- 按图 6-10 所示搭建拓扑；
- 按图 6-10 所示为设备命名并配置 IP 地址；
- 配置合适的路由使得全网连通；
- 配置 PAT 使得内网可以访问公网。

任务实施如下所述。

步骤一：为设备命名并配置接口 IP 地址

（1）在路由器 R1 上为设备命名并配置接口 IP 地址

```
Router(config)#hostname R1
```

```
R1(config)#interface GigabitEthernet0/0
R1(config-if)#ip address 172.16.10.254 255.255.255.0
R1(config-if)#no shutdown
R1(config-if)#interface Serial0/0/0
R1(config-if)#ip address 197.12.16.1 255.255.255.252
R1(config-if)#no shutdown
```

（2）在路由器 ISP 上为设备命名并配置接口 IP 地址

```
Router(config)#hostname ISP
ISP(config)#interface GigabitEthernet0/0
ISP(config-if)#ip address 218.8.8.254 255.255.255.0
ISP(config-if)#no shutdown
ISP(config-if)#interface Serial0/0/0
ISP(config-if)#ip address 197.12.16.2 255.255.255.252
ISP(config-if)#no shutdown
```

步骤二：配置动态 PAT

（1）在路由器 R1 上配置静态默认路由

```
R1(config)# ip route 0.0.0.0 0.0.0.0 197.12.16.2
```

（2）在路由器 R1 上创建 ACL 并标识能够进行转换的地址

```
R1(config)# access-list 1 permit 172.16.10.0 0.0.0.255
```

（3）在路由器 R1 上配置 NAT 地址池（地址池中地址数量只有一个）

```
R1(config)# ip nat pool NAT-POOL 218.8.1.1 218.8.1.1 netmask 255.255.255.255
```

（4）在路由器 R1 上将 ACL 与地址池绑定

```
R1(config)# ip nat inside source list 1 pool NAT-POOL overload
```

（5）在路由器 R1 上配置 NAT 的内部接口与外部接口

```
R1(config)# interface GigabitEthernet 0/0
R1(config-if)#ip nat inside
R1(config)#interface serial 0/0/0
R1(config-if)#ip nat outside
```

（6）在路由器 ISP 上配置到达对端公网 IP 地址的静态主机路由

```
ISP(config)#ip route 218.8.1.1 255.255.255.255 197.12.16.1
```

步骤三：检验动态 PAT 映射

```
R1#show ip nat translations
Pro   Inside global        Inside local      Outside local      Outside global
icmp  218.8.1.1:1024       172.16.10.3:1     218.8.8.100:1      218.8.8.100:1024
icmp  218.8.1.1:1025       172.16.10.4:1     218.8.8.100:1      218.8.8.100:1025
icmp  218.8.1.1:1026       172.16.10.5:1     218.8.8.100:1      218.8.8.100:1026
icmp  218.8.1.1:1          172.16.10.2:1     218.8.8.100:1      218.8.8.100:1
icmp  218.8.1.1:4          172.16.10.1:4     218.8.8.100:4      218.8.8.100:4
icmp  218.8.1.1:5          172.16.10.1:5     218.8.8.100:5      218.8.8.100:5
```

从上面 **show ip nat translations** 命令的输出结果可以看到，网络为不同的内部主机分配了同一个 IP 地址 218.8.1.1（内部全局地址），在 NAT 表中通过不同的端口号来将内部地址区分开来。

```
R1#show ip nat statistics
Total translations: 5 (0 static, 5 dynamic, 5 extended)
Outside Interfaces: Serial0/0/0
Inside Interfaces: GigabitEthernet0/0
Hits: 12    Misses: 15
Expired translations: 9
Dynamic mappings:
-- Inside Source
access-list 1 pool NAT-POOL refCount 5
 pool NAT-POOL: netmask 255.255.255.255
        start 218.8.1.1 end 218.8.1.1
        type generic, total addresses 1 , allocated 1 (100%), misses 0
```

上面 **show ip nat statistics** 输出结果显示，当前共有 5 个动态拓展 NAT 进程正在进行，并且这 5 个地址都被转换成为同一个公网 IP 地址。

步骤四：网络连通性测试

我们以 PC1 为例访问外网的 WEB 服务器，测试内网中的 PC 是否可以访问外网，网络连通性测试结果如图 6-11 所示。

以上测试结果显示：内网中的 PC 可以访问外网。

问题：既然出口地址是一个公网 IP 地址，我们为什么不直接使用接口的公网 IP 地址，还要通过配置一个地址池的方式设置对端路由呢？其实这样做是可以的，这就涉及 PAT 的另外一种实现方法——基于接口的 PAT，请读者依据前面基于端口的 PAT 的语法自行完成配置。

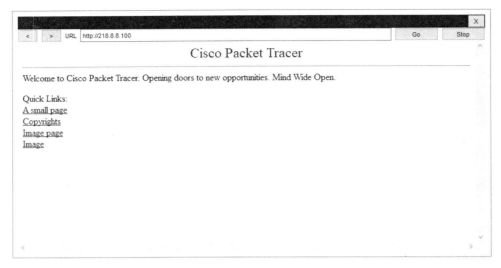

图 6-11　网络连通性测试结果

6.2.5　认识静态 NAT

1. 静态 NAT 的定义

静态 NAT（静态网络地址转换）：将内部本地地址与内部全局地址实现一对一的固定映射。

采用静态 NAT 方式，每一个私网地址都需要占用一个公网地址，且应用后该公网地址不能再应用于其他地址转换。因此静态 NAT 可扩展性极低，主要应用于对外提供服务的内网服务器。

2. 静态 NAT 的语法

Router(config)#**ip nat inside source static** *local-ip global-ip*

- *local-ip*：本地 IP 地址，是用于指定地址映射的内部地址；
- *global-ip*：全局 IP 地址，是用于指定地址映射的外部地址。

Router(config-if)#**ip nat { inside | outside }**

6.2.6　场景三：配置静态 NAT

在内部网络 192.168.10.0/24 中，有 3 台 PC 在访问公网时需要对应地将其 IP 地址转换成 216.9.4.10、216.9.4.20 和 216.9.4.30。通过配置静态 NAT 可实现这种一对一的映射，静态 NAT 实验拓扑如图 6-12 所示。

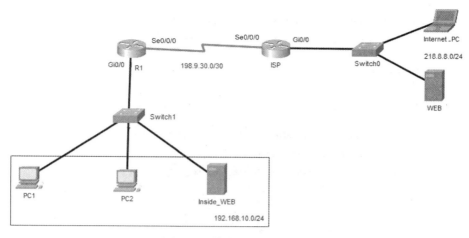

图 6-12 静态 NAT 实验拓扑

任务要求：
- 按图 6-12 所示搭建拓扑；
- 按图 6-12 所示为设备命名并配置 IP 地址；
- 配置合适的路由使得全网连通；
- 配置静态 NAT 使得内网可以访问公网。

任务实施如下所述。

步骤一：为设备命名并配置接口 IP 地址

（1）在路由器 R1 上为设备命名并配置接口 IP 地址

```
Router(config)#hostname R1
R1(config)#interface GigabitEthernet0/0
R1(config-if)#ip address 192.168.10.254 255.255.255.0
R1(config-if)#no shutdown
R1(config-if)#interface Serial0/0/0
R1(config-if)#ip address 198.9.30.1 255.255.255.252
R1(config-if)#no shutdown
```

（2）在路由器 ISP 上为设备命名并配置接口 IP 地址

```
Router(config)#hostname ISP
ISP(config)#interface GigabitEthernet0/0
ISP(config-if)#ip address 218.8.8.254 255.255.255.0
ISP(config-if)#no shutdown
ISP(config-if)#interface Serial0/0/0
```

ISP(config-if)#**ip address 198.9.30.2 255.255.255.252**
ISP(config-if)#**no shutdown**

步骤二：配置静态 NAT

（1）在路由器 R1 上建立内部本地地址与内部全局地址之间的映射关系

R1(config)# **ip nat inside source static 192.168.10.1 216.9.4.10**
R1(config)# **ip nat inside source static 192.168.10.2 216.9.4.20**
R1(config)# **ip nat inside source static 192.168.10.3 216.9.4.30**

（2）在路由器 R1 上配置 NAT 的内部接口和外部接口

R1(config)# **interface GigabitEthernet 0/0**
R1(config-if)#**ip nat inside**
R1(config)#**interface serial 0/0/0**
R1(config-if)#**ip nat outside**

（3）在路由器 ISP 和 R1 上配置到达对端公网 IP 地址的静态路由

ISP(config)# **ip route 216.9.4.0 255.255.255.0 198.9.30.1**
R1(config)#**ip route 0.0.0.0 0.0.0.0 198.9.30.2**

步骤三：查看静态 NAT 映射

R1#**show ip nat translations**

Pro	Inside global	Inside local	Outside local	Outside global
---	216.9.4.10	192.168.10.1	---	---
---	216.9.4.20	192.168.10.2	---	---
---	216.9.4.30	192.168.10.3	---	---

通过以上 **show ip nat translations** 命令的显示结果可以清晰地看到，内部本地地址与内部全局地址之间实现了一对一的映射关系。我们也可以通过 **debug ip nat** 命令查看。

R1#**debug ip nat**

R1#
NAT: s=**192.168.10.1->216.9.4.10**, d=218.8.8.10 [14]
NAT*: s=218.8.8.10, d=216.9.4.10->192.168.10.1 [18]
NAT: s=**192.168.10.2->216.9.4.20**, d=218.8.8.10 [5]
NAT*: s=218.8.8.10, d=216.9.4.20->192.168.10.2 [19]
NAT: s=**192.168.10.3->216.9.4.30**, d=218.8.8.10 [2]

NAT*: s=218.8.8.10, d=216.9.4.30->192.168.10.3 [20]

通过采用 **debug ip nat** 命令进行抓包测试可知：内部本地地址 192.168.10.0/24 网段的 3 台 PC 均根据对应的静态 NAT 映射表访问外网；到达外网 PC 时显示的数据包源 IP 地址为内部全局地址，返回的数据包的目的 IP 地址开始时也是内部全局地址。查看完请务必关闭 **debug** 命令。

```
R1#show ip nat statistics
Total translations: 3 (3 static, 0 dynamic, 0 extended)
Outside Interfaces: Serial0/0/0
Inside Interfaces: GigabitEthernet0/0
Hits: 15    Misses: 21
Expired translations: 19
Dynamic mappings:
```

通过以上 **show ip nat statistics** 命令显示的结果可知：共有 3 条静态映射条目，目前没有进行会话的转换条目。

步骤四：网络连通性测试

与动态 NAT 和 PAT 相比，静态 NAT 要求有足够多的公网 IP 地址。静态 NAT 对于从 Internet 访问内网服务器的操作特别有用，与其他两种 NAT 相比，其映射的固定性有其特殊用途。

下面我们通过使用外网中的 Internet_PC 去访问内网服务器来完成对静态 NAT 的测试。在实验中，我们使用 192.168.10.3 去访问内网服务器，即使用私网 IP 地址访问内网服务器，如图 6-13 所示。

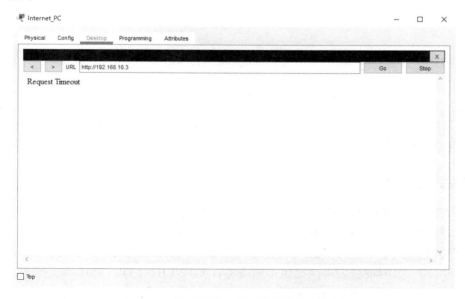

图 6-13　使用私网 IP 地址访问内网服务器

测试结果显示：外网中的 Internet_PC 使用私网 IP 地址 192.168.10.3 直接访问 WEB 服务器失败，即无法访问内网服务器，因为该私网 IP 地址在经过路由器 R1 时进行了一次 NAT，要访问内部服务器，它应该使用内部全局 IP 地址（216.9.4.30），该地址在经过路由器 R1 时会被转换成 192.168.10.3（私网 IP 地址），再去访问内网服务器。

由上可知，路由器 R1 维护了一个 NAT 表，当我们从内网主机访问外网中的服务器时，内网私网地址经过 NAT 转换为公网 IP 地址。同理，当我们使用外网主机访问内网服务器时，也需要使用转换后的公网 IP 地址，这样数据包在进入路由器 R1 后，通过查询 NAT 表找到相应的私网 IP 地址，再去访问内网服务器（仅限静态 NAT）。

使用完成 NAT 操作的公网地址（216.9.4.30）去访问内网服务器，成功访问内网中 Inside_WEB 服务器的界面如图 6-14 所示。

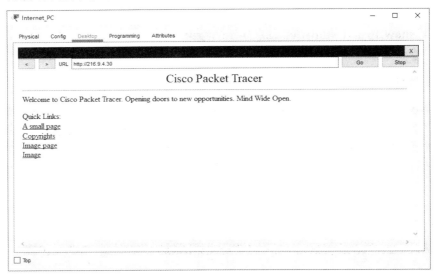

图 6-14　成功使用完成 NAT 操作的公网地址访问内网服务器

6.2.7　认识静态 PAT

1．静态 PAT 的定义

静态 PAT（静态端口地址转换）：将内部本地地址与内部全局地址实现多对一的固定映射，即将固定私网 IP 地址+端口映射为固定公网 IP 地址+端口。

采用静态 PAT 方式，使一个公网地址可以供私网多台主机使用，该方式主要应用于内网多台服务器使用同一公网地址为外网提供不同服务，如 WEB、FTP、SSH 等。公网主机以同一个公网地址访问内网不同的服务，从而提高了公网地址的利用率。

2. 静态 PAT 的语法

Router(config)#**ip nat inside source static {tcp | udp}** *local-ip local-port global-ip global-port*

- **tcp|udp**：协议名称，用来指定要转到端口所属的协议。

Router(config-if)#**ip nat { inside | outside }**

6.2.8 场景四：配置静态 PAT

通过前几节内容的学习，我们已经掌握了 3 种 NAT 的配置，但是 NAT 的功能不仅如此，更为神奇的是，通过一个 IP 地址实现内外网互相访问网服务，通过 Telnet 和 SSH 登录内网网络设备，这就需要用到本节中我们学习的端口映射相关知识——配置静态 PAT。端口映射实验拓扑如图 6-15 所示。

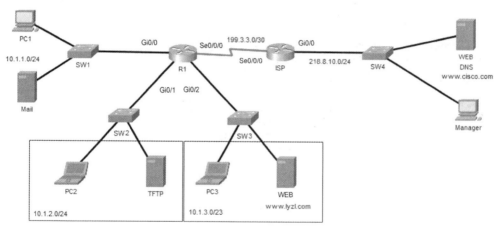

图 6-15 端口映射实验拓扑

任务要求：
- 在 Manager 上采用 Telnet 方式登录交换机 SW1，采用 SSH 方式登录交换机 SW2 和路由器 R1；
- 使 Manager 这台 PC 可以访问 www.lyzl.com 这台 WEB 服务器；
- 使内网中所有的 PC 均可以访问外网 www.cisco.com 这台服务器。

任务实施如下所述。

步骤一：为设备命名并配置接口 IP 地址

（1）在路由器 R1 上为设备命名并配置接口 IP 地址

Router(config)#**hostname R1**

```
R1(config)#interface GigabitEthernet0/0
R1(config-if)#ip address 10.1.1.254 255.255.255.0
R1(config-if)#no shutdown
R1(config)#interface GigabitEthernet0/1
R1(config-if)#ip address 10.1.2.254 255.255.255.0
R1(config-if)#no shutdown
R1(config)#interface GigabitEthernet0/2
R1(config-if)#ip address 10.1.3.254 255.255.255.0
R1(config-if)#no shutdown
R1(config-if)#interface Serial0/0/0
R1(config-if)#ip address 199.3.3.1 255.255.255.252
R1(config-if)#no shutdown
```

（2）在路由器 ISP 上为设备命名并配置接口 IP 地址

```
Router(config)#hostname ISP
ISP(config)#interface GigabitEthernet0/0
ISP(config-if)#ip address 218.8.10.254 255.255.255.0
ISP(config-if)#no shutdown
ISP(config-if)#interface Serial0/0/0
ISP(config-if)#ip address 199.3.3.2 255.255.255.252
ISP(config-if)#no shutdown
```

步骤二：配置动态 PAT

（1）在路由器 R1 上配置命名 ACL

```
R1(config)#ip access-list standard NAT-ACL
R1(config-std-nacl)#permit 10.1.1.0 0.0.0.255
R1(config-std-nacl)#permit 10.1.2.0 0.0.0.255
R1(config-std-nacl)#permit 10.1.3.0 0.0.0.255
R1(config-std-nacl)#deny any
```

（2）在路由器 R1 上配置静态默认路由并配置基于接口的动态 PAT

```
R1(config)#ip route 0.0.0.0 0.0.0.0 199.3.3.2
R1(config)#ip nat inside source list NAT-ACL int s0/0/0 overload
```

（3）在路由器 R1 上配置 NAT 的入接口和出接口

```
R1(config)# interface GigabitEthernet 0/0
```

```
R1(config-if)#ip nat inside
R1(config)# interface GigabitEthernet 0/1
R1(config-if)#ip nat inside
R1(config)# interface GigabitEthernet 0/2
R1(config-if)#ip nat inside
R1(config)#interface Serial 0/0/0
R1(config-if)#ip nat outside
```

步骤三：配置 Telnet 和 SSH

（1）在交换机 SW1 上配置 Telnet

```
SW1(config)#interface vlan 1
SW1(config-if)#ip address 10.1.1.200 255.255.255.0
SW1(config-if)#no shutdown
SW1(config-if)#exit
SW1(config)#ip default-gateway 10.1.1.254
SW1(config)#line vty 0 3
SW1(config-line)#password cisco
SW1(config-line)#login
SW1(config-line)#exit
SW1(config)#enable secret ytvc
SW1(config)#service password-encryption
```

（2）在交换机 SW2 上配置 SSH

```
SW2(config)#interface vlan 1
SW2(config-if)#ip address 10.1.2.200 255.255.255.0
SW2(config-if)#no shutdown
SW2(config-if)#exit
SW2(config)#ip default-gateway 10.1.2.254
SW2(config)#username ZJQ password cisco
SW2(config)#ip domain-name ytvc.com
SW2(config)#line vty 0 3
SW2(config-line)#transport input ssh
SW2(config-line)#login local
SW2(config-line)#exit
SW2(config)#enable secret ytvc
SW2(config)#crypto key generate rsa
```

The name for the keys will be: **SW2.ytvc.com**
Choose the size of the key modulus in the range of 360 to 2048 for your General Purpose Keys. Choosing a key modulus greater than 512 may take a few minutes.

How many bits in the modulus [512]: **1024**
% Generating 1024 bit RSA keys, keys will be non-exportable...[OK]

SW2(config)#**service password-encryption**

（3）在路由器 R1 上配置 SSH

R1(config)#**username CFL password cisco**
R1(config)#**ip domain-name ytvc.com**
R1(config)#**line vty 0 3**
R1(config-line)#**transport input ssh**
R1(config-line)#**login local**
R1(config-line)#**exit**
R1(config)#**enable secret ytvc**
R1(config)#**crypto key generate rsa**
The name for the keys will be: **R1.ytvc.com**
Choose the size of the key modulus in the range of 360 to 2048 for your General Purpose Keys. Choosing a key modulus greater than 512 may take a few minutes.

How many bits in the modulus [512]: **1024**
% Generating 1024 bit RSA keys, keys will be non-exportable...[OK]

R1(config)#**service password-encryption**

步骤四：配置静态 PAT

R1(config)#**ip nat inside source static tcp 10.1.3.100 80 199.3.3.1 80**
R1(config)#**ip nat inside source static tcp 10.1.1.200 23 199.3.3.1 23**
R1(config)#**ip nat inside source static tcp 10.1.2.200 22 199.3.3.1 22**
R1(config)#**ip nat inside source static tcp 10.1.3.254 22 199.3.3.1 22**

不难看出，在配置端口映射时需要熟记一些常见协议端口号。例如，SSH 的端口号是 22、HTTP 的端口号是 80。

步骤五：配置 DNS 服务器

众所周知，用域名访问 WEB 服务器，首先需要将域名解析为 IP 地址，这就要求先配置

DNS 服务器，如图 6-16 所示。

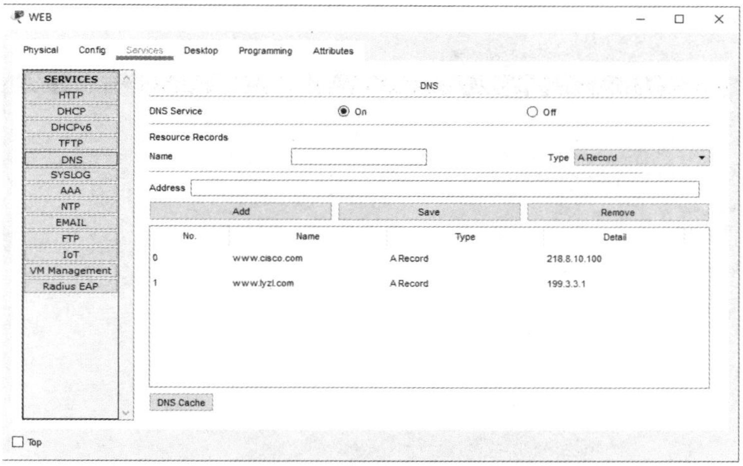

图 6-16　配置 DNS 服务器

配置 DNS 服务器只需要将域名与 IP 地址相对应即可。需要注意的是，内网 WEB 服务器的域名将被解析为完成 NAT 操作的 IP 地址，这和采用静态 NAT 方式访问公网时的情况大致相同。

步骤六：PAT 验证测试

如图 6-17 所示，访问外网网站，该图展示了内网主机 PC3 成功访问外网中 www.cisco.com 这台 WEB 服务器，动态 PAT 配置成功。

图 6-17　访问外网网站

如图 6-18 所示，外网主机 Manager 成功访问内网服务器，静态 PAT 配置成功。

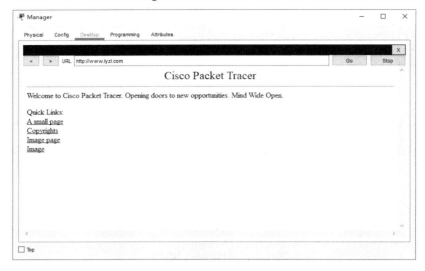

图 6-18　访问内网服务器

在外网 PC 上通过 Telnet 方式远程登录交换机 SW1：

C:\>**telnet 199.3.3.1**

Trying 199.3.3.1 ...Open

User Access Verification

Password:
SW1>

在外网 PC 上通过 SSH 方式远程登录交换机 SW2：

C:\>**ssh -l ZJQ 199.3.3.1**

Password:

SW2>**enable**
Password:
SW2#

在内网 PC 上通过 SSH 方式远程登录路由器 R1：

```
C:\>ssh -l CFL 199.3.3.1

Password:




R1>enable
Password:
R1#
```

6.2.9 常见 NAT 故障排错命令

NAT 常见故障排错命令如下：
- 使用 **show ip nat translations** 命令可检验转换表是否包含正确的转换条目。
- 使用 **clear** 和 **debug** 命令可以清除 IPv4 动态创建的 NAT 表项（静态配置的 NAT 表项一直存储在 NAT 表中），查看 IPv4 NAT 的过程。
- 使用 **show ip nat statistics** 命令可显示活动转换条目的总数、NAT 配置参数、地址池中地址数量和已分配地址数量等信息，用于查看 IPv4 NAT 的统计信息。
- 使用 **debug ip nat** 命令可查看 IPv4 NAT 的过程，显示有关 NAT 数据包的信息以检验 NAT 功能。
- 使用 **show access-list** 命令可以检验将要转换哪些内部地址，以及 ACL 是否配置正确。

NAT 技术对于 IPv4 地址紧缺现状确实有一定的缓解作用，却不是长久之计，要想彻底解决地址紧缺问题，需要使用 IPv6 地址。

6.3 配置拓展 NAT

6.3.1 场景五：在无线路由器上完成 NAT 内置配置

有心的读者也许已经发现，在前面介绍的内容中，我们都是通过命令直接在路由器进行相关配置的，下面我们将学习在无线路由器上完成 NAT 内置配置。内置 NAT 配置实验拓扑如图 6-19 所示。

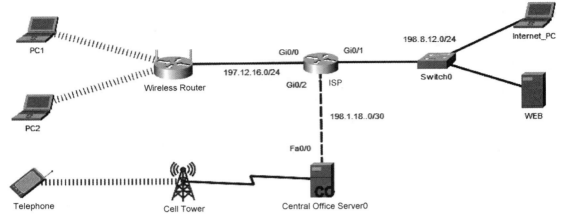

图 6-19 内置 NAT 配置实验拓扑

任务要求：
- 按图 6-19 所示搭建拓扑；
- 按图 6-19 所示为设备命名并配置 IP 地址；
- 配置内置 NAT 使得内网可以访问公网。

任务实施如下所述。

步骤一：为设备命名并配置接口 IP 地址

在路由器 ISP 上为设备命名并配置接口 IP 地址：

```
Router(config)#hostname ISP
ISP(config)#interface GigabitEthernet0/0
ISP(config-if)#ip address 197.12.16.2 255.255.255.252
ISP(config-if)#no shutdown
ISP(config)#interface GigabitEthernet0/1
ISP(config-if)#ip address 198.8.12.254 255.255.255.0
ISP(config-if)#no shutdown
ISP(config)#interface GigabitEthernet0/2
ISP(config-if)#ip address 198.1.18.2 255.255.255.252
ISP(config-if)#no shutdown
```

步骤二：配置 NAT 过载

（1）在 Wireless Router 上配置 NAT 过载

在 Wireless Router 上配置 NAT 过载，如图 6-20 所示。

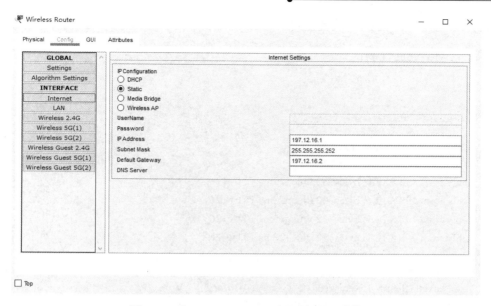

图 6-20　在 Wireless Router 上配置 NAT 过载

（2）在服务器 Central Office Server0 上配置内置 NAT

在服务器 Central Office Server0 上配置内置 NAT，如图 6-21 所示。

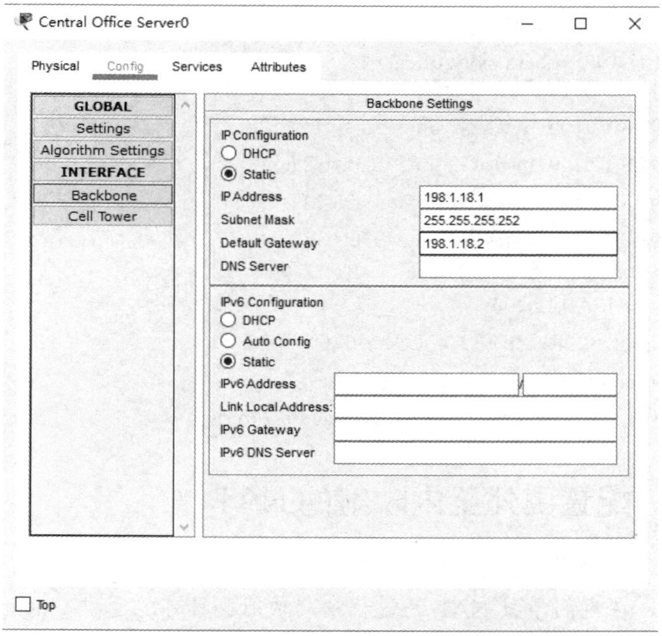

图 6-21　在服务器 Central Office Server0 上配置内置 NAT

注意：在配置内置 NAT 时网关是对端的 IP 地址。

步骤三：网络连通性测试

（1）无线网络测试

```
C:\>ping 198.8.12.1

Pinging 198.8.12.1 with 32 bytes of data:

Reply from 198.8.12.1: bytes=32 time=23ms TTL=126
Reply from 198.8.12.1: bytes=32 time=28ms TTL=126
Reply from 198.8.12.1: bytes=32 time=19ms TTL=126
Reply from 198.8.12.1: bytes=32 time=28ms TTL=126

Ping statistics for 198.8.12.1:
    Packets: Sent = 4, Received = 4, Lost = 0 (0% loss),
Approximate round trip times in milli-seconds:
    Minimum = 19ms, Maximum = 28ms, Average = 24ms
```

（2）3G/4G 网络测试

```
C:\>ping 198.8.12.100

Pinging 198.8.12.100 with 32 bytes of data:

Reply from 198.8.12.100: bytes=32 time=5ms TTL=126
Reply from 198.8.12.100: bytes=32 time=25ms TTL=126
Reply from 198.8.12.100: bytes=32 time=39ms TTL=126
Reply from 198.8.12.100: bytes=32 time=33ms TTL=126

Ping statistics for 198.8.12.100:
    Packets: Sent = 4, Received = 4, Lost = 0 (0% loss),
Approximate round trip times in milli-seconds:
    Minimum = 5ms, Maximum = 39ms, Average = 25ms
```

6.3.2　场景六：配置由外至内映射的 NAT

前面我们已经学习了动态 NAT、静态 NAT 和 PAT，相信大家已经掌握了这 3 种 NAT 的配置。细心的朋友可能发现，在我们学习 NAT 的过程中，映射条目可以通过 Outside 创建，接下来，让我们以静态 NAT 为例，一起来完成 Inside 映射实验。Inside 映射实验拓扑如图 6-22 所示。

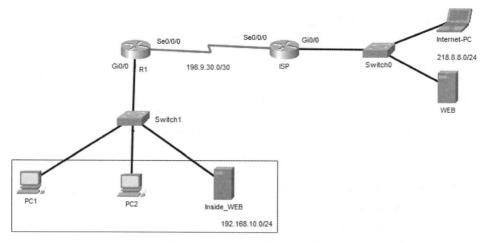

图 6-22　Inside 映射实验拓扑

任务要求：
- 按图 6-22 所示搭建拓扑；
- 按图 6-22 所示为设备命名并配置 IP 地址；
- 配置合适的路由使得全网连通；
- 配置 Inside 映射使得内网可以访问公网。

任务实施如下所述。

步骤一：为设备命名并配置接口 IP 地址

（1）在路由器 R1 上为设备命名并配置接口 IP 地址

```
Router(config)#hostname R1
R1(config)#interface GigabitEthernet0/0
R1(config-if)#ip address 192.168.10.254 255.255.255.0
R1(config-if)#no shutdown
R1(config-if)#interface Serial0/0/0
R1(config-if)#ip address 198.9.30.1 255.255.255.252
R1(config-if)#no shutdown
```

（2）在路由器 ISP 上为设备命名并配置接口 IP 地址

```
Router(config)#hostname ISP
ISP(config)#interface GigabitEthernet0/0
ISP(config-if)#ip address 218.8.8.254 255.255.255.0
ISP(config-if)#no shutdown
ISP(config-if)#interface Serial0/0/0
```

ISP(config-if)#**ip address 198.9.30.2 255.255.255.252**
ISP(config-if)#**no shutdown**

步骤二：配置 NAT

（1）在路由器 R1 上建立内部本地地址与内部全局地址之间的映射关系

R1(config)# **ip nat inside source static 192.168.10.1 216.9.4.10**

（2）在路由器 R1 上配置 NAT 的内部接口与外部接口

R1(config)# **interface GigabitEthernet 0/0**
R1(config-if)#**ip nat inside**
R1(config)#**interface Serial 0/0/0**
R1(config-if)#**ip nat outside**

（3）在路由器 ISP 和 R1 上配置到达对端的静态路由

ISP(config)# **ip route 192.168.10.0 255.255.255.0 198.9.30.1**
R1(config)#**ip route 0.0.0.0 0.0.0.0 198.9.30.2**

步骤三：查看静态 NAT 映射

R1#**show ip nat translations**
Pro	Inside global	Inside local	Outside local	Outside global
---	---	---	192.168.10.254	198.9.30.1

通过 **show ip nat translations** 命令可以清晰地查看到，外部本地地址与外部全局地址之间一对一映射。

R1#show ip nat statistics
Total translations: 1 (0 static, 1 dynamic, 0 extended)
Outside Interfaces: Serial0/0/0
Inside Interfaces: GigabitEthernet0/0
Hits: 0   Misses: 34
Expired translations: 0
Dynamic mappings:

show ip nat statistics 命令的输出结果显示，当前只有 1 个 NAT 进程正在进行。

步骤四：网络连通性测试

我们以 Internet_PC 访问 Inside_WEB 为例来完成网络连通性测试，如图 6-23 所示，由该图可知，Internet_PC 可以访问 Inside_WEB。

图 6-23 网络连通性测试

6.4 挑战练习

6.4.1 挑战练习一

挑战要求：CFL 老师和 16NET1 班的同学们共同设计了一个实验拓扑，准备考一考 16NET2 班的同学们，一是为了帮助同学们巩固刚学过的 NAT 知识，二是为了让同学们发现自己的不足。挑战练习一网络拓扑和地址规划表分别如图 6-24 和表 6-1 所示。

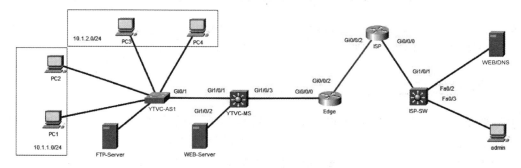

图 6-24 挑战练习一网络拓扑

表 6-1　挑战练习一地址规划表

设备名称	接口	IP 地址	子网掩码	接口描述 / 默认网关
YTVC-AS1	VLAN 100	10.1.0.250	255.255.255.0	10.1.0.254
YTVC-MS	VLAN 10	10.1.1.254	255.255.255.0	——
	VLAN 20	10.1.2.254	255.255.255.0	——
	VLAN 100	10.1.0.254	255.255.255.0	——
	Gi1/0/2	10.0.0.254	255.255.255.0	Link to WEB-Server
	Gi1/0/3	10.0.100.1	255.255.255.252	Link to Edge Gi0/0/0
Edge	Gi0/0/0	10.0.100.2	255.255.255.252	Link to YTVC-MS Gi1/0/3
	Gi0/0/2	217.12.7.2	255.255.255.252	Link to ISP Gi0/0/2
ISP	Gi0/0/0	218.6.26.254	255.255.255.0	Link to ISP-SW Gi1/0/1
	Gi0/0/2	217.12.7.1	255.255.255.252	Link to Edge Gi0/0/2
FTP-Server	NIC	10.1.0.250	255.255.255.0	10.1.0.254
		——	——	Link to YTVC-AS-1 Gi0/2
WEB-Server	NIC	10.0.0.250	255.255.255.0	10.0.0.254
		——	——	Link to YTVC-MS Gi1/0/2
WEB/DNS	NIC	218.6.26.200	255.255.255.0	218.6.26.254
		——	——	Link to ISP-SW Gi1/0/3

任务要求：

- 要求内网设备均支持远程管理，其中三层交换机 YTVC-MS 采用 SSH 方式登录，交换机 YTVC-AS1 和路由器 Edge 采用 Telnet 方式登录；
- 要求外网主机 admin 可以访问内网服务器 WEB-Server（其域名为 www.ytvc.com.cn 和 www.ytvc.edu.cn）；
- 要求内网所有 PC 均可以访问外网服务器 WEB/DNS（其域名为 www.cisco.com）；
- 要求将所有网络设备配置文件备份到内网服务器 FTP-Server 上。

6.4.2　挑战练习二

挑战要求：CFL 老师将 16NET1 班设计的图 6-24 所示的网络拓扑交给 16NET2 班的同学们后，16NET2 班的同学们决定礼尚往来，于是也和 CFL 老师一起设计了一个网络拓扑——挑战练习二实验拓扑，要以此来考一考 16NET1 班的同学们。挑战练习二网络拓扑和地址规划表分别如图 6-25 和表 6-2 所示，该图中所有 PC 均通过 DHCP 获取 IP 地址。

第 6 章 学习 NAT 技术 <<< 227

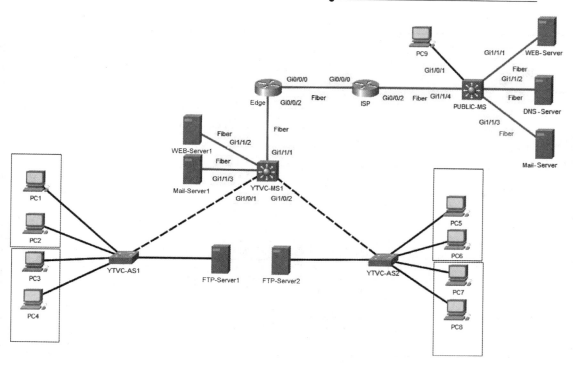

图 6-25 挑战练习二网络拓扑

表 6-2 挑战练习二地址规划表

设备名称	接口	IP 地址	子网掩码	接口描述/默认网关
YTVC-AS1	VLAN 99	10.0.99.1	255.255.255.0	10.0.99.254
YTVC-AS2	VLAN 99	10.0.99.2	255.255.255.0	10.0.99.254
YTVC-MS1	VLAN 10	10.1.1.254	255.255.255.0	——
	VLAN 20	10.1.2.254	255.255.255.0	——
	VLAN 30	10.2.1.254	255.255.255.0	——
	VLAN 40	10.2.2.254	255.255.255.0	——
	VLAN 99	10.0.99.254	255.255.255.0	——
	VLAN 100	10.1.0.254	255.255.255.0	——
	VLAN 200	10.2.0.254	255.255.255.0	——
	Gi1/1/1	10.0.0.1	255.255.255.252	Link to Edge Gi0/0/2
	Gi1/1/2	10.0.1.254	255.255.255.0	Link to WEB-Server1
	Gi1/1/3	10.0.2.254	255.255.255.0	Link to Mail-Server1
Edge	Gi0/0/2	10.0.0.2	255.255.255.252	Link to YTVC-MS1 Gi1/1/1
	Gi0/0/0	217.12.8.5	255.255.255.252	Link to ISP Gi0/0/0

续表

设备名称	接口	IP 地址	子网掩码	接口描述/默认网关
ISP	Gi0/0/0	217.12.8.6	255.255.255.252	Link to Edge Gi0/0/0
	Gi0/0/2	218.9.1.254	255.255.255.0	Link to PUBLIC-MS Gi1/1/4
FTP-Server1	NIC	10.1.0.253	255.255.255.0	10.1.0.254
		—	—	Link to YTVC-AS1 Gi0/2
FTP-Server2	NIC	10.2.0.253	255.255.255.0	10.2.0.254
		—	—	Link to YTVC-AS2 Gi0/2
WEB-Server1	NIC	10.0.1.253	255.255.255.0	10.0.1.254
		—	—	Link to YTVC-MS1 G1/1/2
Mail-Server1	NIC	10.0.2.253	255.255.255.0	10.0.2.254
		—	—	Link to YTVC-MS1 Gi1/1/3
WEB-Server	NIC	218.9.1.200	255.255.255.0	218.9.1.254
				Link to PUBLIC-MS Gi1/1/1
DNS-Server	NIC	218.9.1.100	255.255.255.0	218.9.1.254
				Link to PUBLIC-MS G1/1/2
Mail-Server	NIC	218.9.1.10	255.255.255.0	218.9.1.254
				Link to PUBLIC-MS Gi1/1/3

任务要求：

- 要求内网中的所有网络设备均可以实现远程访问，其中 YTVC-MS1 使用 SSH 方式，YTVC-AS1 使用 Telnet 方式；
- 使用同一个 IP 地址实现内部网络访问外部网络的 WEB-Server，同时外部网络也可访问内网中的 WEB-Server1（要求使用域名访问）；
- 配置 NTP 服务器，实现网络拓扑时间统一；
- 将所有的网络设备配置文件备份到 FTP-Server1 和 FTP-Server2 上；
- 已知 YTVC-MS1 配置文件被管理员误删除，请你借助 FTP-Server1 和 FTP-Server2 将 YTVC-MS1 配置恢复。

6.5 本章小结

本章内容到此结束。本章主要内容包括认识 NAT 技术、配置多种类型的 NAT 和配置拓展 NAT。NAT 技术是有效缓解 IPv4 地址短缺的重要手段，可控制内网主机对外网的访问，同时也可控制外网主机对内网的访问，同时有效避免了来自外网的网络攻击，在很大程度上提高了网络安全性。本章设计了 6 个应用场景和 2 个挑战练习，使读者能够理解 NAT 的应用场景，在现实网络中熟练应用 NAT 技术。

第 7 章 >>>

学习基础 IPv6

本章要点:

- 学习 IPv6 基础知识
- 配置 IPv6 地址
- 配置 SLAAC 与 DHCPv6
- 挑战练习
- 本章小结

IPv6 技术彻底解决了 IPv4 地址短缺的问题。本章介绍多种类型 IPv6 地址的配置，为高级 IPv6 的学习奠定基础。其中，7.1 节介绍 IPv6 地址类型，以及 ICMPv6 和 NDP；7.2 节介绍多种类型 IPv6 地址的配置，包括 IPv6 全球单播地址、EUI-64 地址、链路本地地址和任播地址；7.3 节介绍 SLAAC（无状态地址自动配置），以及无状态 DHCPv6 和状态化 DHCPv6。本章设计了 7 个应用场景和 2 个挑战练习，使读者能够熟练配置 IPv6 地址，做到活学活用。

7.1 学习 IPv6 基础知识

7.1.1 学习 IPv6 地址

为解决 IPv4 地址枯竭问题，IPv6 技术应运而生了。相比 IPv4 地址的 32 位长度，IPv6 地址的长度增加到了 128 位，提供足够大的地址空间，这意味着 IP 地址将不再短缺，可以说地球上的每一粒沙子都可以分配到一个 IPv6 地址。IPv6 地址与 IPv4 地址比较如表 7-1 所示。

表 7-1 IPv6 地址与 IPv4 地址比较

地址类型			IPv6 地址	IPv4 地址
单播地址	全局单播地址	可聚合全局单播地址	2000::/3	IPv4 公网地址
	本地单播地址	链路本地地址	FE80::/10	169.254.0.0/16 自动专用 IP 地址
		站点本地地址	FEC0::/10	10.0.0.0/8、172.16.0.0/12、192.168.0.0/16
	特殊单播地址	不确定地址	::/128	0.0.0.0
		环回地址	::1/128	127.0.0.1～127.255.255.254
组播地址			FF00::/8	224.0.0.0/4
任播地址			在单播地址空间中分配	无
广播地址			无	主机位全为 1

注释：在 IPv6 地址中，主机 ID 全为 1 和全为 0 的地址可以分配给 IPv6 的主机。

说明：

① 全局单播地址：具有全局唯一性，是 Internet 可路由地址，可静态配置也可以动态分配。

② 链路本地地址：仅限于单个链路，链路指子网，链路之外具有不可路由性。

③ 站点本地地址：仅限于有限站点间的本地编址，在全局 IPv6 上不可路由。

④ 组播地址：标识一组接口，仅可用于目的地址，IPv6 中没有广播地址，广播功能通过组播实现。主要组播地址有节点组播地址 FF02::1、路由器组播地址 FF02::2、OSPF 路由器组播地址 FF02::5、OSPF DR 路由器组播地址 FF02::6。

⑤ 任播地址：在单播地址空间中分配，仅用作目标地址，限于子网内，只分配给路由器。

注意：一台 IPv6 主机的一个接口可以支持多个 IPv6 地址，这些 IPv6 地址可以同属于一个子网也可以属于不同的子网。

7.1.2 认识 ICMPv6

与 IPv4 一样，IPv6 也需要一个控制协议用于在 IP 主机、路由器之间传递控制消息，实现 ICMP（Internet Control Message Protocol，互联网控制报文协议）的功能。由于 ICMP 不能满足 IPv6 的全部要求，因此人们开发了新版本的 ICMP，称为 ICMPv6（Internet Control Message Protocol Version 6，第 6 版互联网控制报文协议），ICMPv6 是为了与 IPv6 配套使用而开发的。ICMPv6 的功能与 ICMP 相似，但在 ICMPv6 中许多消息的意义与在 ICMP 中并不相同，例如，源抑制和时间戳消息。在 IPv4 中，ICMP 使用的协议号为 1。在 IPv6 中，ICMPv6 使用的"下一个报头"字段的值为 58。ICMPv6 可向源节点报告关于目的地址传输 IPv6 数据包的错误和信息，具有差错报告、网络诊断、邻节点发现和组播实现等功能。在 IPv6 中，ICMPv6 可实现 IPv4 中 ICMP、ARP 和 IGMP 的功能。ICMPv6 报文基本格式如表 7-2 所示，各字段含义说明如下。

表 7-2　ICMPv6 报文的基本格式

类型（1 字节）	代码（1 字节）	校验和（2 字节）
ICMP 报文体（可变长度）		

- 类型：标识 ICMPv6 报文类型，它的值根据报文的内容来确定；
- 代码：用于进一步确定 ICMPv6 信息，对同一类型的报文进行了更详细的分类；
- 校验和：用于检测 ICMPv6 报文是否被正确传送；
- 报文体：返回出错的参数，记录出错报文的片段，帮助源节点判断错误原因或参数。

7.1.3 认识 NDP

NDP（Neighbor Discovery Protocol，邻居发现协议）是 TCP/IP 协议栈的一部分，主要应用于 IPv6。NDP 工作在数据链路层，其主要作用是负责在链路上发现其他节点和相应的地址，确定可用路由，维护可用路径和其他活动节点信息的可达性。

NDP 通过 ICMPv6 报文来承载，在一个 IPv6 数据报中，如果该数据报的"下一个报头"字段的值为 58，且 ICMPv6 报文中类型字段取值范围为 133～137，则此 IPv6 报文的数据部分含有 NDP 报文。NDP 定义了以下 5 种 ICMPv6 报文类型。

- 类型 133：路由器请求（RS），由主机发起，用于请求路由器发送一个 RA 消息。
- 类型 134：路由器通告（RA），由路由器发起，通告路由器的存在和链路的细节参数（链路前缀、MTU、跳数限制等）。该消息被周期性发送，也用于应答 RS。
- 类型 135：邻居请求（NS），由节点主机发起，用于请求另一台主机的数据链路层地址，或实现地址冲突检测、邻居不可达检测。

- 类型 136：邻居通告（NA），由节点发起，用于响应 NS。如果一个节点改变了数据链路层地址，则它会主动发送一个 NA 通告新地址。
- 类型 137：重定向。

NDP 使用 ICMPv6 报文实现地址解析、邻居不可达检测、地址冲突检测、地址自动配置、路由器发现、前缀发现以及重定向等功能。NDP 消息通常在链路本地范围内收发。

7.2 配置 IPv6 地址

7.2.1 场景一：配置 IPv6 全局单播地址

通过前面 7.1 节内容的学习，我们已经对 IPv6 地址有所了解，接下来，让我们通过具体实验来配置 IPv6 全局单播地址。配置 IPv6 全局单播地址实验拓扑如图 7-1 所示。

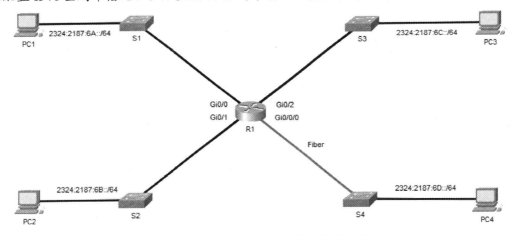

图 7-1　配置 IPv6 全局单播地址实验拓扑

任务要求：
- 按图 7-1 所示搭建拓扑；
- 按图 7-1 所示为设备命名并配置 IPv6 地址；
- 配置全局单播地址实现全网连通。

任务实施如下所述。

步骤一：配置设备 IPv6 全局单播地址

（1）在路由器 R1 上配置接口 IPv6 地址

R1(config)#**ipv6 unicast-routing**　　//开启 IPv6 路由转发功能

```
R1(config)#interface GigabitEthernet0/0
R1(config-if)#ipv6 address 2324:2187:6A::/64
R1(config-if)#no shutdown
R1(config-if)#interface GigabitEthernet0/1
R1(config-if)#ipv6 address 2324:2187:6B::/64
R1(config-if)#no shutdown
R1(config-if)#interface GigabitEthernet0/2
R1(config-if)#ipv6 address 2324:2187:6C::/64
R1(config-if)#no shutdown
R1(config-if)#interface GigabitEthernet0/0/0
R1(config-if)#ipv6 address 2324:2187:6D::/64
R1(config-if)#no shutdown
```

（2）在客户端 PC1 上配置 IPv6 地址

如图 7-2 所示，在客户端 PC1 上配置 IPv6 地址。

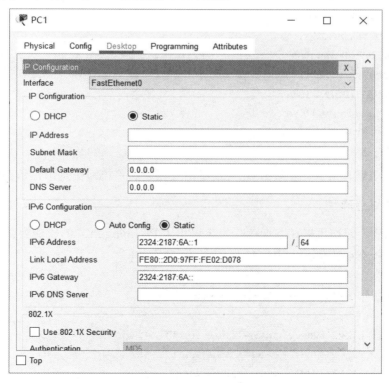

图 7-2　在客户端 PC1 上配置 IPv6 地址

步骤二：查看 IPv6 路由表

```
R1#show ipv6 route
IPv6 Routing Table - 9 entries
<output omitted>
C 2324:2187:6A::/64 [0/0]
    via GigabitEthernet0/0, directly connected
L 2324:2187:6A::/128 [0/0]
    via GigabitEthernet0/0, receive
C 2324:2187:6B::/64 [0/0]
    via GigabitEthernet0/1, directly connected
L 2324:2187:6B::/128 [0/0]
    via GigabitEthernet0/1, receive
C 2324:2187:6C::/64 [0/0]
    via GigabitEthernet0/2, directly connected
L 2324:2187:6C::/128 [0/0]
    via GigabitEthernet0/2, receive
C 2324:2187:6D::/64 [0/0]
    via GigabitEthernet0/0/0, directly connected
L 2324:2187:6D::/128 [0/0]
    via GigabitEthernet0/0/0, receive
L FF00::/8 [0/0]
    via Null0, receive
```

由以上路由表输出可知，4 条 IPv6 直连路由及相应的主机路由全部显示在路由表中。

步骤三：测试连通性

此时，所有 PC 间均可以相互通信。以下展示了 PC1 ping PC4 的成功测试结果。

```
C:\>ping 2324:2187:6D::1

Pinging 2324:2187:6D::1 with 32 bytes of data:

Reply from 2324:2187:6D::1: bytes=32 time=1ms TTL=127
Reply from 2324:2187:6D::1: bytes=32 time<1ms TTL=127
Reply from 2324:2187:6D::1: bytes=32 time<1ms TTL=127
Reply from 2324:2187:6D::1: bytes=32 time<1ms TTL=127
```

Ping statistics for 2324:2187:6D::1:
Packets: Sent = 4, Received = 4, Lost = 0 (0% loss),
Approximate round trip times in milli-seconds:
Minimum = 0ms, Maximum = 1ms, Average = 0ms

至此，IPv6 全局单播地址配置实验已经结束。需要读者注意的是，一个 IPv6 接口可以拥有多个 IPv6 全局单播地址，所以配置新的 IPv6 地址不会覆盖之前已配置的 IPv6 地址。

7.2.2 场景二：配置 IPv6 的 EUI-64 地址

EUI-64（64-bit Extended Unique Identifier，64 位扩展唯一标识符）是由IEEE定义的地址。IEEE EUI-64 地址代表网络接口寻址的新标准，提供了一种自动配置 IPv6 主机地址的方法。IPv6 设备将使用其接口的 MAC 地址来生成唯一的 64 位接口 ID，根据接口的 48 位 MAC 地址创建 EUI-64 格式地址。首先将 MAC 地址分为两个 24 位组，第一个 24 位组是 OUI（Organizationally Unique Identifier，组织唯一标识符），第二个 24 位组用于表示特定的 NIC（Network Interface Card，以太网网卡），然后将 16 位的 FFFE 插入这两个 24 位组之间以形成 64 位 EUI-64 地址。

我们已经了解 EUI-64 地址的生成方法，接下来，让我们一起来完成该特殊单播地址的 EUI-64 地址配置，配置 EUI-64 地址实验拓扑如图 7-3 所示。

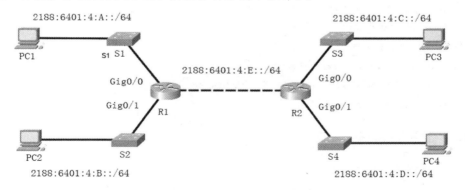

图 7-3 配置 EUI-64 地址实验拓扑

任务要求：
● 按图 7-3 所示搭建拓扑；
● 按图 7-3 所示配置路由器接口的 IPv6 EUI-64 地址；
● 参考 8.1 节内容完成 IPv6 静态路由配置，实现全网互通。
任务实施如下所述。

步骤一：配置设备接口 IPv6 EUI-64 地址

（1）配置路由器 R1 接口 EUI-64 地址

```
R1(config)#ipv6 unicast-routing
R1(config)#interface GigabitEthernet 0/0
R1(config-if)#ipv6 address 2188:6401:4:A::/64 eui-64
R1(config-if)#no shutdown
R1(config-if)#interface GigabitEthernet 0/1
R1(config-if)#ipv6 address 2188:6401:4:B::/64 eui-64
R1(config-if)#no shutdown
R1(config-if)#interface GigabitEthernet 0/2
R1(config-if)#ipv6 address 2188:6401:4:E::/64 eui-64
R1(config-if)#no shutdown
```

（2）配置路由器 R2 接口 EUI-64 地址

```
R2(config)#ipv6 unicast-routing
R2(config)#interface GigabitEthernet 0/0
R2(config-if)#ipv6 address 2188:6401:4:C::/64 eui-64
R2(config-if)#no shutdown
R2(config-if)#interface GigabitEthernet 0/1
R2(config-if)#ipv6 address 2188:6401:4:D::/64 eui-64
R2(config-if)#no shutdown
R2(config-if)#interface GigabitEthernet 0/2
R2(config-if)#ipv6 address 2188:6401:4:E::/64 eui-64
R2(config-if)#no shutdown
```

步骤二：查看 IPv6 接口配置

（1）查看路由器 R1 接口 EUI-64 地址

```
R1>show ipv6 interface brief
GigabitEthernet0/0        [up/up]
    FE80::209:7CFF:FEA7:B45C
    2188:6401:4:A:209:7CFF:FEA7:B45C      //EUI-64 地址
GigabitEthernet0/1        [up/up]
    FE80::201:63FF:FE38:BA8B
    2188:6401:4:B:201:63FF:FE38:BA8B      //EUI-64 地址
GigabitEthernet0/2        [up/up]
```

```
            FE80::20A:41FF:FEAB:EBEB
            2188:6401:4:E:20A:41FF:FEAB:EBEB    //EUI-64 地址，作为步骤三中路由器 R2 的下一跳
```

（2）查看路由器 R2 接口 EUI-64 地址

```
    R2>show ipv6 interface brief
    GigabitEthernet0/0          [up/up]
        FE80::201:64FF:FE64:DAC1
        2188:6401:4:C:201:64FF:FE64:DAC1        //EUI-64 地址
    GigabitEthernet0/1          [up/up]
        FE80::2D0:D3FF:FE77:D808
        2188:6401:4:D:2D0:D3FF:FE77:D808        //EUI-64 地址
    GigabitEthernet0/2          [up/up]
        FE80::2D0:BAFF:FE2C:48D6
        2188:6401:4:E:2D0:BAFF:FE2C:48D6        //EUI-64 地址，作为步骤三中路由器 R1 的下一跳
```

步骤三：配置 IPv6 静态路由

（1）配置路由器 R1 静态路由

```
    R1(config)#ipv6 route 2188:6401:4:C::/64 2188:6401:4:E:2D0:BAFF:FE2C:48D6
    R1(config)#ipv6 route 2188:6401:4:D::/64 2188:6401:4:E:2D0:BAFF:FE2C:48D6
```

（2）配置路由器 R2 静态路由

```
    R2(config)#ipv6 route 2188:6401:4:A::/64 2188:6401:4:E:20A:41FF:FEAB:EBEB
    R2(config)#ipv6 route 2188:6401:4:B::/64 2188:6401:4:E:20A:41FF:FEAB:EBEB
```

步骤四：测试网络连通性

此时，所有 PC 间均可以相互通信。以下展示了 PC1 跟踪 PC4 的成功测试结果。

```
    C:\>tracert 2188:6401:4:D::1
    Tracing route to 2188:6401:4:D::1 over a maximum of 30 hops:
        1   0ms     0ms     0ms     2188:6401:4:A:209:7CFF:FEA7:B45C    //R1EUI-64 地址
        2   0ms     0ms     0ms     2188:6401:4:E:2D0:BAFF:FE2C:48D6    //R2EUI-64 地址
        3   0ms     0ms     0ms     2188:6401:4:D::1                    //目的主机 PC4
    Trace complete.
```

至此，配置 EUI-64 地址实验已经结束，需要读者注意的是，EUI-64 也是单播地址的一种，只不过 64 位的接口 ID 是自动生成的。

7.2.3 场景三：配置 IPv6 链路本地地址

在配置完 IPv6 全局单播地址和 EUI-64 地址之后，相信读者会意犹未尽，接下来，让我们通过具体实验一起来配置 IPv6 链路本地地址。配置 IPv6 链路本地地址实验拓扑如图 7-4 所示。

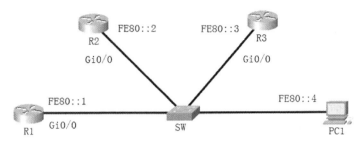

图 7-4　配置 IPv6 链路本地地址实验拓扑

任务要求：
- 按图 7-4 所示搭建拓扑；
- 按图 7-4 所示为设备配置 IPv6 链路本地地址；
- 所有设备通过链路本地地址实现互通。

任务实施如下所述。

步骤一：配置设备接口 IPv6 链路本地地址

（1）配置路由器 R1 接口 IPv6 链路本地地址

```
R1(config)#ipv6 unicast-routing
R1(config)#interface GigabitEthernet0/0
R1(config-if)#ipv6 address fe80::1 link-local
R1(config-if)#no shutdown
```

（2）配置路由器 R2 接口 IPv6 链路本地地址

```
R2(config)#ipv6 unicast-routing
R2(config)#interface GigabitEthernet0/0
R2(config-if)#ipv6 address fe80::2 link-local
R2(config-if)#no shutdown
```

（3）配置路由器 R3 接口 IPv6 链路本地地址

```
R3(config)#ipv6 unicast-routing
```

R3(config)#**interface GigabitEthernet0/0**

R3(config-if)#**ipv6 address fe80::3 link-local**

R3(config-if)#**no shutdown**

（4）配置 PC1 IPv6 链路本地地址

通过图形化界面将 PC1 的 "Link Local Address" 静态地址指定为 FE80::4。

步骤二：查看路由表信息

先使用 **show ipv6 interface brief** 命令依次查看 3 台路由器的链路本地地址，确保配置成功。接下来，使用 **show ipv6 route** 命令依次查看 3 台路由器的路由表，请思考路由表中为什么没有链路本地地址的直连路由。

步骤三：测试设备连通性

此时，网络中所有设备均可以通过链路本地地址实现通信。以下展示了路由器 R1 与其余 3 台设备间的连通性测试结果。

R1>**ping fe80::2**

Output Interface: **GigabitEthernet0/0**

Type escape sequence to abort.

Sending 5, 100-byte ICMP Echos to FE80::2, timeout is 2 seconds:

!!!!!

Success rate is 100 percent (5/5), round-trip min/avg/max = 0/0/0ms

以上输出表明，路由器 R1 与路由器 R2 通信成功。

R1>**ping fe80::3**

Output Interface: **GigabitEthernet0/0**

Type escape sequence to abort.

Sending 5, 100-byte ICMP Echos to FE80::3, timeout is 2 seconds:

!!!!!

Success rate is 100 percent (5/5), round-trip min/avg/max = 0/0/0ms

以上输出表明，路由器 R1 与路由器 R3 通信成功。

R1>**ping fe80::4**

Output Interface: **GigabitEthernet0/0**

Type escape sequence to abort.

Sending 5, 100-byte ICMP Echos to FE80::4, timeout is 2 seconds:

!!!!!

Success rate is 100 percent (5/5), round-trip min/avg/max = 0/0/0ms

以上输出表明，路由器 R1 与 PC1 通信成功。

请读者在 PC1 上测试 PC1 与 3 台路由器的连通性，进一步体会链路本地地址的作用。请尝试删除链路本地地址。

7.2.4 场景四：配置 IPv6 任播地址

任播可为互联网上特定类型的网络服务提供冗余和负载功能。任播地址的实质是为同一服务的多台设备分配一个共享全局地址，这些设备位于网络中的不同站点。发送到任播地址的数据包被发送给此地址所标识的一组接口中距离源节点路由最近的一个接口。

任播技术被广泛应用于大型 DNS 部署，通过配置任播地址可改善一些 DNS 服务器因链路故障、设备故障等导致的通信阻断状况，从而增强了 DNS 服务器的生存能力。

通过以上介绍，我们已经大体了解了任播的概念，为尽快掌握任播知识，接下来，让我们一起来配置 IPv6 任播地址。配置 IPv6 任播地址实验拓扑如图 7-5 所示。

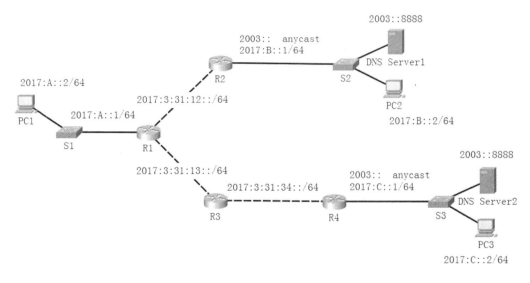

图 7-5　配置 IPv6 任播地址实验拓扑

任务需求：

- 按图 7-5 所示搭建网络拓扑；
- 按图 7-5 所示为设备配置 IPv6 地址；
- 配置 IPv6 任播地址，使 2 台 DNS Server 同时为 PC1 提供服务；
- 使用主机 PC2 和 PC3 模拟 WEB 服务器（其域名为 www.yttv.com），假设主机 PC2 和 PC3 数据相同；

- 当主机 PC1 访问 WEB 服务器（域名为 www.yttv.com）时，首选 DNS Server1，如果 DNS Server1 无法访问则选择备用服务器 DNS Server2；
- 参考 8.2.6 节内容完成 OSPFv3 动态路由协议的配置，实现全网互通。

任务实施如下所述：

步骤一：配置设备接口 IPv6 地址

（1）配置路由器 R1 接口 IPv6 地址

```
R1(config)#ipv6 unicast-routing
R1(config)#interface fastEthernet 0/0
R1(config-if)#ipv6 address 2017:A::1/64
R1(config-if)#no shutdown
R1(config-if)#interface fastEthernet 0/1
R1(config-if)#ipv6 address 2017:3:31:12::1/64
R1(config-if)#no shutdown
R1(config-if)#interface fastEthernet 1/0
R1(config-if)#ipv6 address 2017:3:31:13::1/64
R1(config-if)#no shutdown
```

（2）配置路由器 R2 接口 IPv6 地址

```
R2(config)#ipv6 unicast-routing
R2(config)#interface fastEthernet 0/1
R2(config-if)#ipv6 address 2017:3:31:12::2/64
R2(config-if)#no shutdown
R2(config-if)#interface fastEthernet 0/0
R2(config-if)#ipv6 address 2003::/64 anycast    //路由器 R2 接口地址 2003::/64 为任播地址
R2(config-if)#ipv6 address 2017:B::1/64
R2(config-if)#no shutdown
```

（3）配置路由器 R3 接口 IPv6 地址

```
R3(config)#ipv6 unicast-routing
R3(config)#interface fastEthernet 0/1
R3(config-if)#ipv6 address 2017:3:31:13::3/64
R3(config-if)#no shutdown
R3(config)#interface fastEthernet 0/0
R3(config-if)#ipv6 address 2017:3:31:34::3/64
R3(config-if)#no shutdown
```

（4）配置路由器 R4 接口 IPv6 地址

```
R4(config)#ipv6 unicast-routing
R4(config)#interface fastEthernet 0/0
R4(config-if)#ipv6 address 2017:3:31:34::4/64
R4(config-if)#no shutdown
R4(config)#interface fastEthernet 0/1
R4(config-if)#ipv6 address 2003::/64 anycast        //路由器 R4 接口地址 2003::/64 为任播地址
R4(config-if)#ipv6 address 2017:C::1/64
R4(config-if)#no shutdown
```

（5）配置相关终端设备

配置 PC1、PC2、PC3 的 IPv6 地址分别为 2017:A::2/64、2017:B::2/64、2017:C::2/64，同时为其配置正确的网关，DNS 地址为 2003:8888。

① 在 DNS Server1 上配置 IPv6 地址，如图 7-6 所示。

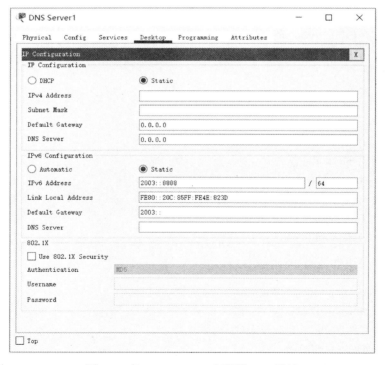

图 7-6　在 DNS Server1 上配置 IPv6 地址

② 在 DNS Server1 上配置 DNS 服务，如图 7-7 所示。

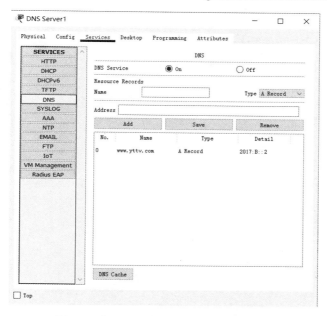

图 7-7　在 DNS Server1 上配置 DNS 服务

③ 在 DNS Server2 上配置 IPv6 地址，如图 7-8 所示。

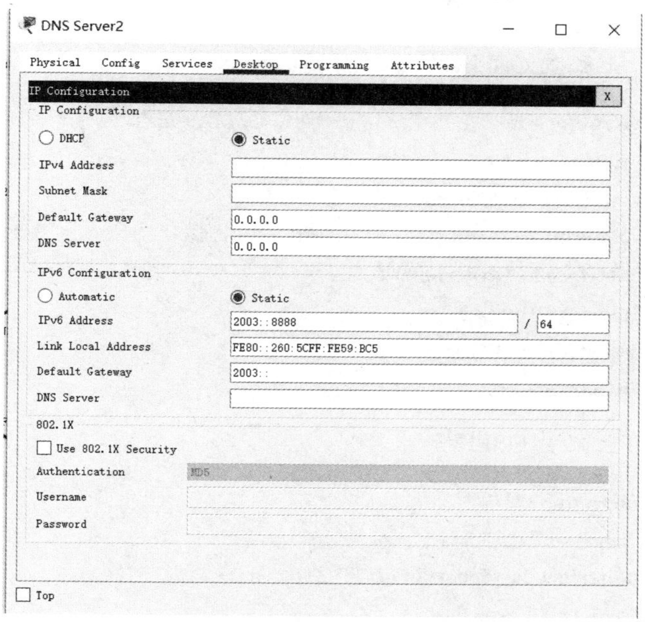

图 7-8　在 DNS Server2 上配置 IPv6 地址

④ 在 DNS Server2 上配置 DNS 服务，如图 7-9 所示。

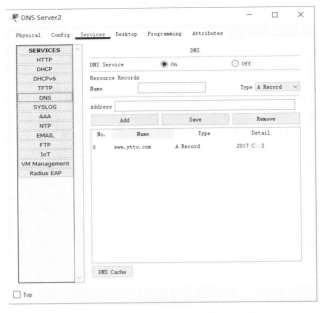

图 7-9 在 DNS Server2 上配置 DNS 服务

步骤二：配置动态路由协议 OSPFv3

（1）在路由器 R1 上配置 OSPFv3

R1(config)#**ipv6 router ospf 1**	//创建 OSPFv3 进程
R1(config-rtr)#**router-id 1.1.1.1**	
R1(config-rtr)#**interface fastEthernet 0/0**	
R1(config-if)#**ipv6 ospf 1 area 0**	//接口调用进程 1 并加入区域 0
R1(config-if)#**interface fastEthernet 0/1**	
R1(config-if)#**ipv6 ospf 1 area 0**	
R1(config-if)#**interface fastEthernet 1/0**	
R1(config-if)#**ipv6 ospf 1 area 0**	

（2）在路由器 R2 上配置 OSPFv3

```
R2(config)#ipv6 router ospf 1
R2(config-rtr)#router-id 2.2.2.2
R2(config-rtr)#interface fastEthernet 0/1
R2(config-if)#ipv6 ospf 1 area 0
R2(config-if)#interface fastEthernet 0/0
R2(config-if)#ipv6 ospf 1 area 0
```

（3）在路由器 R3 上配置 OSPFv3

```
R3(config)#ipv6 router ospf   1
R3(config-rtr)#router-id 3.3.3.3
R3(config-rtr)#interface fastEthernet 0/1
R3(config-if)#ipv6 ospf 1 area 0
R3(config-if)#interface fastEthernet 0/0
R3(config-if)#ipv6 ospf 1 area 0
```

（4）在路由器 R4 上配置 OSPFv3

```
R4(config)#ipv6 router ospf   1
R4(config-rtr)#router-id 4.4.4.4
R4(config-rtr)#interface fastEthernet 0/0
R4(config-if)#ipv6 ospf 1 area 0
R4(config-if)#interface fastEthernet 0/1
R4(config-if)#ipv6 ospf 1 area 0
```

步骤三：查看 IPv6 路由表

```
R1#show ipv6 route
IPv6 Routing Table - 11 entries
<output omitted>
O  2003::/64 [110/2]
   via FE80::2D0:D3FF:FEC3:DD02, FastEthernet0/1      //下一跳为 Fa0/1 接口连接的 R2
C  2017:3:31:12::/64 [0/0]
   via FastEthernet0/1, directly connected
L  2017:3:31:12::1/128 [0/0]
   via FastEthernet0/1, receive
C  2017:3:31:13::/64 [0/0]
   via FastEthernet1/0, directly connected
L  2017:3:31:13::1/128 [0/0]
   via FastEthernet1/0, receive
O  2017:3:31:34::/64 [110/2]
   via FE80::2D0:97FF:FE24:C902, FastEthernet1/0
C  2017:A::/64 [0/0]
   via FastEthernet0/0, directly connected
L  2017:A::1/128 [0/0]
   via FastEthernet0/0, receive
O  2017:B::/64 [110/2]
```

via FE80::2D0:D3FF:FEC3:DD02, FastEthernet0/1
O 2017:C::/64 [110/3]
via FE80::2D0:97FF:FE24:C902, FastEthernet1/0
L FF00::/8 [0/0]
via Null0, receive

步骤四：完成访问服务器测试

C:\>tracert www.yttv.com
Tracing route to 2017:B::2 over a maximum of 30 hops:

1	0ms	0ms	0ms	2017:A::1	//R1
2	0ms	0ms	0ms	2017:3:31:12::2	//R2
3	0ms	0ms	0ms	2017:B::2	//www.yttv.com

Trace complete.

步骤五：模拟故障并观察路径切换情况

（1）将路由器 R2 接口 Fa 0/0 关闭后查看路由表

R1#show ipv6 route
IPv6 Routing Table - **8 entries**
<output omitted>
O 2003::/64 [110/3]
via FE80::2D0:97FF:FE24:C902, FastEthernet1/0   //下一跳切换为 Fa1/0 接口连接的 R3
C 2017:3:31:13::/64 [0/0]
via FastEthernet1/0, directly connected
L 2017:3:31:13::1/128 [0/0]
via FastEthernet1/0, receive
O 2017:3:31:34::/64 [110/2]
via FE80::2D0:97FF:FE24:C902, FastEthernet1/0
C 2017:A::/64 [0/0]
via FastEthernet0/0, directly connected
L 2017:A::1/128 [0/0]
via FastEthernet0/0, receive
O 2017:C::/64 [110/3]
via FE80::2D0:97FF:FE24:C902, FastEthernet1/0
L FF00::/8 [0/0]
via Null0, receive

观察发现，路径被切换到"备选路径"，再用 PC1 访问 WEB 服务器（其域名为 www.yttv.com）：

```
C:\>tracert www.yttv.com
Tracing route to 2017:C::2 over a maximum of 30 hops:
  1    0ms      0ms      0ms      2017:A::1            //R1
  2    0ms      0ms      0ms      2017:3:31:13::3      //R3
  3    0ms      0ms      0ms      2017:3:31:34::4      //R4
  4    0ms      0ms      0ms      2017:C::2            //www.yttv.com
Trace complete.
```

（2）将路由器 R2 接口 Fa 0/0 开启后，等 OSPFv3 收敛后再次查看选路

此时，我们可以看到，路径又被切换回以路由器 R2 为下一跳的"最优路径"，实验结束。需要注意的是，迄今为止，任播技术的定义依然不十分清楚，本案例基于已知资料设计并得到了 PT 支持。本书使用的 PT 8.0 及以前版本还不能完美支持任播技术。

7.3 配置 SLAAC 与 DHCPv6

7.3.1 认识 SLAAC 与 DHCPv6

SLAAC（Stateless Address Auto-Configuration，无状态地址自动配置）可在没有 DHCPv6 服务器情况下，使客户端通过 SLAAC 提供的服务自动构建 IPv6 全局单播地址，自动获取默认网关地址。SLAAC 是一种无状态服务，因此不能维护网络中的 IPv6 地址数据库。

SLAAC 核心是 ICMPv6，通过 ICMPv6 的 RA 消息（Router Advertisement，路由器通告）可提供 DHCP 服务功能，客户端根据 RA 消息自动构建 IPv6 地址。RA 消息由 IPv6 路由器每 200 s 发送一次。SLAAC 仅由路由器通告 RA，RA 管理标志位（M）和其他标志位（O）均被置零。SLAAC 应用于不关心主机在网络中使用的具体地址，只求地址唯一可路由的网络环境。

DHCPv6 是为 IPv6 网络上的主机自动分配 IPv6 地址、网络前缀、前缀长度、域名等配置参数（除了默认网关地址）的动态主机配置协议。DHCPv6 分为状态化 DHCPv6 和无状态 DHCPv6 两种。状态化 DHCPv6 与 DHCPv4 功能类似，都会去维护客户端的状态信息；而无状态 DHCPv6 是 SLAAC 和 DHCPv6 应用的组合。在 IPv6 网络环境中，学会配置 DHCPv6，可以大大简化网络管理员的工作。

7.3.2 学习 DHCPv6 配置语法

DHCPv6 配置语法总结如下：

（1）服务器端主要配置

```
Server(config)#ipv6 unicast-routing                    //启用 IPv6 路由功能
```

```
Server(config)#ipv6 dhcp pool name                              //定义 DHCPv6 地址池名称
Server(config-dhcpv6)#dns-server x.x.x.x                        //定义 DNS Server 的地址
Server(config-dhcpv6)#domain-name name                          //定义域名
Server(config-if)#ipv6 nd { managed-config-flag | other-config-flag }   //接口启用状态化 / 无状态 DHCPv6
Server(config-if)#ipv6 dhcp server name                         //接口应用 DHCPv6 地址池
```

（2）客户端主要配置

```
Client(config)#ipv6 unicast-routing
Client(config-if)#ipv6 enable                                   //允许路由器接收该链路发送的 RS 消息和参与 DHCPv6
Client(config-if)#ipv6 address { autoconfig | dhcp }            //定义获取 IPv6 地址的方式
```

注释：SLAAC 只需要开启 IPv6 路由功能即可。

7.3.3 场景五：完成 SLAAC（无状态地址自动配置）

在了解 SLAAC 之后，相信很多读者一定想知道，如何在没有 DHCPv6 服务器的情况下，使客户端获取 IPv6 地址。接下来，让我们通过具体实验，一起来完成 SLAAC，验证客户端是否可以获取 IPv6 地址。SLAAC 实验拓扑如图 7-10 所示。

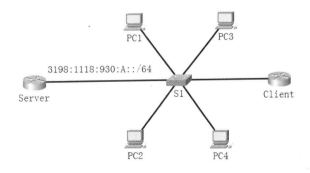

图 7-10　SLAAC 实验拓扑

任务要求：
- 按图 7-10 所示搭建拓扑；
- 按图 7-10 所示完成 SLAAC（无状态地址自动配置）；
- 网络中所有设备可以通过 IPv6 地址进行通信。

任务实施如下所述。

步骤一：配置路由器 Server 接口 IPv6 地址

```
Server(config)#interface GigabitEthernet0/0/0
Server(config-if)#ipv6 address Fe80::1 link-local               //配置接口链路本地地址，可省略
```

```
Server(config-if)#ipv6 address 3198:1118:930:A::/64
Server(config-if)#no shutdown
```

步骤二：配置 SLAAC 服务器

```
Server(config)#ipv6 unicast-routing              //SLAAC 只需要开启 IPv6 路由功能即可
Server(config)#interface GigabitEthernet0/0/0
Server(config-if)#no ipv6 nd managed-config-flag //将 RA 消息中 M 标记置 0（可省略，默认命令）
Server(config-if)#no ipv6 nd other-config-flag   //将 RA 消息中 O 标记置 0（可省略，默认命令）
```

步骤三：配置 SLAAC 客户端

（1）在路由器 Client 上配置 SLAAC 客户端

```
Client(config)#ipv6 unicast-routing
Client(config)#interface GigabitEthernet0/0/0
Client(config-if)#no shutdown
Client(config-if)#ipv6 enable                    //允许路由器接收该链路发送 RS 消息和启用 DHCPv6
Client(config-if)#ipv6 address autoconfig        //将路由器接口启用为 DHCPv6 客户端
```

（2）在终端 PC 上配置 SLAAC 客户端

如图 7-11 所示，在 PC 上配置无状态 DHCPv6 客户端，请选择"Automatic"选项。

图 7-11　在 PC 上配置无状态 DHCPv6 客户端

步骤四：查看 IPv6 地址获取情况

（1）在路由器 Client 上查看 IPv6 地址获取情况

```
Client>show ipv6 route
IPv6 Routing Table - 4 entries
Codes: C - Connected, L - Local, S - Static, R - RIP, B - BGP
       U - Per-user Static route, M - MIPv6
       I1 - ISIS L1, I2 - ISIS L2, IA - ISIS interarea, IS - ISIS summary
       ND - ND Default, NDp - ND Prefix, DCE - Destination, NDr - Redirect
       O - OSPF intra, OI - OSPF inter, OE1 - OSPF ext 1, OE2 - OSPF ext 2
       ON1 - OSPF NSSA ext 1, ON2 - OSPF NSSA ext 2
       D - EIGRP, EX - EIGRP external
ND   ::/0 [2/0]
     via FE80::1, GigabitEthernet0/0/0             //通过 SLAAC 服务获取 IPv6 默认网关
NDp 3198:1118:930:A::/64 [2/0]
     via GigabitEthernet0/0/0, directly connected
L    3198:1118:930:A:201:43FF:FEEA:1154/128 [0/0]  //构建的全局单播地址
     via GigabitEthernet0/0/0, receive
L    FF00::/8 [0/0]
     via Null0, receive
```

（2）在 PC 客户端上查看 IPv6 地址获取情况

```
C:\>ipv6config /all
FastEthernet0 Connection:(default port)           //查看主机 PC1 网卡的 IPv6 地址
    Connection-specific DNS Suffix…… :
    Physical Address............................: 0050.0FBE.B0BE
    Link-local IPv6 Address....................: FE80::250:FFF:FEBE:B0BE
    IPv6 Address..................................:3198:1118:930:A:250:FFF:FEBE:B0BE   //构建全局单播地址
    Default Gateway............................: FE80::1         //通过 SLAAC 服务获取 IPv6 默认网关
    DNS Servers...................................: ::
    DHCPv6 IAID.................................: 1529611506
    DHCPv6 Client DUID........................: 00-01-00-01-C9-D6-1C-A3-00-50-0F-BE-B0-BE
```

步骤五：测试设备间连通性

此时所有设备均可相互通信。以下展示了路由器 Server 和路由器 Client（客户端）的成功测试结果。

```
Server>ping 3198:1118:930:A:201:43FF:FEEA:1154
```

Sending 5, 100-byte ICMP Echos to 3198:1118:930:A:201:43FF:FEEA:1154, timeout is 2 seconds:
!!!!!
Success rate is 100 percent (5/5), round-trip min/avg/max = 0/0/0ms

以下展示了路由器 Server 和客户端 PC1 的成功测试结果。

Server>ping **3198:1118:930:A:250:FFF:FEBE:B0BE**
Sending 5, 100-byte ICMP Echos to 3198:1118:930:A:250:FFF:FEBE:B0BE, timeout is 2 seconds:
!!!!!
Success rate is 100 percent (5/5), round-trip min/avg/max = 0/0/0ms

至此，SLAAC 实验结束。所谓 SLAAC 就是在没有 DHCPv6 服务器情况下，由路由器为 IPv6 客户端提供 IPv6 前缀及默认网关。

7.3.4 场景六：配置无状态 DHCPv6 服务

在了解 DHCPv6 服务之后，我们知道了 DHCPv6 服务能够自动给客户端分配 IPv6 地址，不再需要人为配置。因此学会配置 DHCPv6 服务将会简化我们的工作。接下来，让我们通过具体实验一起完成无状态 DHCPv6 服务配置。无状态 DHCPv6 配置实验拓扑如图 7-12 所示。

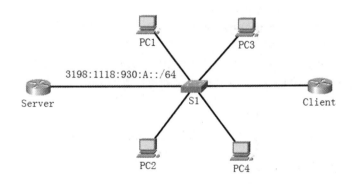

图 7-12　无状态 DHCPv6 配置实验拓扑

任务要求：
- 按图 7-12 所示搭建拓扑；
- 按图 7-12 所示配置无状态 DHCPv6 服务；
- 网络中所有设备均可以通过 IPv6 地址进行通信。

任务实施如下所述。

步骤一：配置路由器 Server 接口 IPv6 地址

Server(config)#**interface GigabitEthernet0/0/0**

```
Server(config-if)#ipv6 address Fe80::255 link-local    //配置接口链路本地地址，可省略
Server(config-if)#ipv6 address 3198:1118:930:A::/64
Server(config-if)#no shutdown
```

步骤二：配置路由器无状态 DHCPv6 服务器

```
Server(config)#ipv6 unicast-routing
Server(config)#ipv6 dhcp pool LAN-IPv6                 //创建 DHCPv6 地址池
Server(config-dhcpv6)#dns-server 2021::8888
Server(config-dhcpv6)#domain-name www.ytvc
Server(config-dhcpv6)#interface GigabitEthernet0/0/0
Server(config-if)#ipv6 nd other-config-flag            //将 RA 消息中 O 标记置 1

Server(config-if)#ipv6 dhcp server LAN-IPv6            //将接口与 DHCPv6 地址池绑定
```

步骤三：配置无状态 DHCPv6 客户端

（1）在路由器 Client 上配置无状态 DHCPv6 客户端

```
Client(config)#ipv6 unicast-routing
Client(config)#interface GigabitEthernet0/0/0
Client(config-if)#no shutdown
Client(config-if)#ipv6 enable           //允许路由器接收该链路发送的 RS 和启用 DHCPv6
Client(config-if)#ipv6 address autoconfig    //将路由器接口启用为 DHCPv6 客户端
```

（2）在终端 PC 上配置无状态 DHCPv6 客户端

PC 的无状态 DHCPv6 客户端配置，请选择"Automatic"选项。

步骤四：查看 IPv6 地址获取情况

（1）在路由器 Client 上查看 IPv6 地址获取情况

```
Client>show ipv6 route
IPv6 Routing Table - 4 entries
Codes: C - Connected, L - Local, S - Static, R - RIP, B - BGP
       U - Per-user Static route, M - MIPv6
       I1 - ISIS L1, I2 - ISIS L2, IA - ISIS interarea, IS - ISIS summary
       ND - ND Default, NDp - ND Prefix, DCE - Destination, NDr - Redirect
       O - OSPF intra, OI - OSPF inter, OE1 - OSPF ext 1, OE2 - OSPF ext 2
       ON1 - OSPF NSSA ext 1, ON2 - OSPF NSSA ext 2
       D - EIGRP, EX - EIGRP external
```

```
ND      ::/0 [2/0]
         via FE80::255, GigabitEthernet0/0/0         //通过 SLAAC 服务获取到 IPv6 默认网关
NDp 3198:1118:930:A::/64 [2/0]
         via GigabitEthernet0/0/0, directly connected
L       3198:1118:930:A:201:43FF:FEEA:1154/128 [0/0]  //构建全局单播地址
         via GigabitEthernet0/0/0, receive
L       FF00::/8 [0/0]
         via Null0, receive
```

此时，Client 通过 DHCPv6 服务自动获取的 DNS 已显示在运行配置文件中，如下所示。

```
Client#show running-config | include server
ip name-server 2021::8888
```

（2）在 PC 客户端上查看 IPv6 地址获取情况

```
C:\>ipv6config /all
FastEthernet0 Connection:(default port)              //查看主机 PC1 网卡的 IPv6 地址参数
   Connection-specific DNS Suffix.....: www.ytvc      //通过 DHCP 服务器获取的域名
   Physical Address..................: 0050.0FBE.B0BE
   Link-local IPv6 Address............: FE80::250:FFF:FEBE:B0BE
   IPv6 Address......................:3198:1118:930:A:250:FFF:FEBE:B0BE //构建全局单播地址
   Default Gateway...................: FE80::255       //通过 SLAAC 服务获取到 IPv6 默认网关
   DNS Servers.......................: 2021::8888      //通过 DHCP 服务器获取 DNS
   DHCPv6 IAID......................: 1805894554
   DHCPv6 Client DUID...............: 00-01-00-01-C9-D6-1C-A3-00-50-0F-BE-B0-BE
```

步骤五：测试设备间连通性

此时，所有设备均可相互通信，请读者自行完成测试。

无状态 DHCPv6 服务实际是 SLAAC 和 DHCPv6 应用的组合，两者相互补充分配参数。客户端可通过 SLAAC 服务获取 IPv6 网络前缀、前缀长度以及默认网关 IPv6 地址，自动构建 IPv6 全局单播地址；DHCPv6 服务器提供 SLAAC，没有提供其他配置信息（DNS 和域名等）。

7.3.5　场景七：配置状态化 DHCPv6 服务

经过 7.3.4 节的学习，我们已经掌握了无状态 DHCPv6 服务的配置，接下来，让我们通过实验一起来完成状态化 DHCPv6 配置。状态化 DHCPv6 配置实验拓扑如图 7-13 所示。

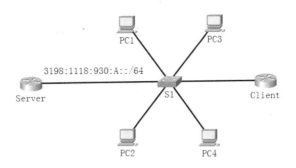

图 7-13 状态化 DHCPv6 配置实验拓扑

任务要求：
- 按图 7-13 所示搭建拓扑；
- 按图 7-13 所示配置状态化 DHCPv6 服务；
- 网络中所有设备均可以通过 IPv6 地址进行通信。

任务实施如下所述。

步骤一：配置路由器 Server 接口 IPv6 地址

```
Server(config)#interface GigabitEthernet0/0/0
Server(config-if)#ipv6 address Fe80::731 link-local    //配置接口链路本地地址，可省略
Server(config-if)#ipv6 address 3198:1118:930:A::/64
Server(config-if)#no shutdown
```

步骤二：配置路由器状态化 DHCPv6 服务器

在路由器 Server 上配置有状态 DHCPv6：

```
Server(config)#ipv6 unicast-routing
Server(config)#ipv6 dhcp pool Stateful-DHCPv6-LAN      //创建 DHCPv6 地址池
Server(config-dhcpv6)#address prefix 3198:1118:930:A::/64
Server(config-dhcpv6)#dns-server 2021::7316
Server(config-dhcpv6)#domain-name ytvc.edu
Server(config-dhcpv6)#interface GigabitEthernet0/0/0
Server(config-if)#ipv6 nd managed-config-flag          //将 RA 消息中 M 标记置 1
Server(config-if)#ipv6 dhcp server Stateful-DHCPv6-LAN //将接口与 DHCPv6 地址池绑定
```

步骤三：配置状态化 DHCPv6 客户端

（1）在路由器 Client 上配置状态化 DHCPv6 客户端

```
Client(config)#ipv6 unicast-routing
Client(config)#interface GigabitEthernet0/0/0
```

```
Client(config-if)#no shutdown
Client(config-if)#ipv6 enable            //允许路由器接收该链路发送的 RS 消息和启用 DHCPv6
Client(config-if)#ipv6 address dhcp      //将路由器接口启用为 DHCPv6 客户端
```

（2）在终端 PC 上配置状态化 DHCPv6 客户端

PC 的状态化 DHCPv6 客户端配置，请选择"Automatic"选项。

步骤四：查看 IPv6 地址获取情况

（1）在路由器 Client 上查看 IPv6 地址获取情况

```
Client>show ipv6 route
IPv6 Routing Table - 4 entries
<output omitted>
ND    ::/0 [2/0]
      via FE80::731, GigabitEthernet0/0/0              //通过 SLAAC 服务获取 IPv6 默认网关
C     3198:1118:930:A::/64 [0/0]
      via GigabitEthernet0/0/0, directly connected
L     3198:1118:930:A:6A9A:6A9A:6A9A:6A9A/128 [0/0]    //通过 DHCPv6 自动获取 IPv6 地址
      via GigabitEthernet0/0/0, receive
L     FF00::/8 [0/0]
      via Null0, receive
```

此时，Client 通过 DHCPv6 服务自动获取的 DNS 显示在运行配置文件中，如下所示。

```
Client#show running-config | include server
ip name-server 2021::7316
```

（2）在 PC 客户端上查看 IPv6 地址获取情况

```
C:\>ipv6config /all
FastEthernet0 Connection:(default port)               //查看主机 PC2 网卡的 IPv6 地址参数
    Connection-specific DNS Suffix..  : ytvc.edu      //通过 DHCPv6 服务器获取的域名
    Physical Address................  : 00D0.978B.17E6
    Link-local IPv6 Address.........  : FE80::2D0:97FF:FE8B:17E6
    IPv6 Address....................  :3198:1118:930:A:999F:999F:999F:999F    //通过 DHCPv6 自
动获取 IPv6 地址
    Default Gateway.................  : FE80::731     //通过 SLAAC 服务获取 IPv6 默认网关
    DNS Servers.....................  : 2021::7316    //通过 DHCPv6 服务器获取的 DNS
    DHCPv6 IAID.....................  : 9113889
    DHCPv6 Client DUID..............  : 00-01-00-01-8A-A5-BB-67-00-D0-97-8B-17-E6
```

步骤五：测试设备间连通性

此时所有设备均可相互通信，请读者自行完成测试。以下展示了客户端 Client 与 PC2 的连通性测试结果。

```
Client#ping 3198:1118:930:A:999F:999F:999F:999F

Sending 5, 100-byte ICMP Echos to 3198:1118:930:A:999F:999F:999F:999F, timeout is 2 seconds:
!!!!!
Success rate is 100 percent (5/5), round-trip min/avg/max = 0/0/0ms
```

状态化 DHCPv6 配置实验证明，网络中需要 DHCPv6 服务器，客户端仅从 SLAAC 服务的 RA 消息中获取默认网关 IPv6 地址，而从 DHCPv6 服务器获取剩余 IPv6 配置参数，如 IPv6 地址、网络前缀、前缀长度、DNS 及域名等。

7.4 挑战练习

7.4.1 挑战练习一

挑战要求：通过学习本章的内容，我们已经掌握了配置 IPv6 地址的方法，包括静态指定和动态分配 2 种方法。细心的读者可能会发现，本章没有涉及 IPv6 站点本地地址配置，接下来，让我们通过一个趣味挑战去配置所有类型的 IPv6 地址。配置 IPv6 站点本地地址实验拓扑如图 7-14 所示。

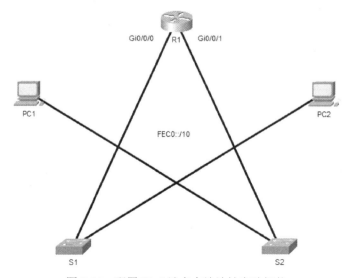

图 7-14　配置 IPv6 站点本地地址实验拓扑

任务要求：
- 按图 7-14 所示搭建拓扑；
- 按图 7-14 所示为设备配置 IP 地址；
- 为路由器每个接口手动指定配置 4 个全局单播地址（一个网段或多个网段）、1 个 EUI-64 地址、1 个站点本地地址和链路本地地址；
- 配置状态化 DHCPv6，要求 PC 获取的是站点本地地址，确保 PC 互通；
- 查看路由器 IPv6 路由表，分析路由表中有多少条记录，缺少哪种类型地址；
- 查看路由器接口信息，按顺序逐个删除接口 IPv6 地址，删除后立即检验。

7.4.2 挑战练习二

挑战要求：经过学习本章的内容，相信读者已经掌握了动态获取 IPv6 地址的方法。细心的读者可能会发现，在前面的案例场景中，都是用路由器作为 DHCPv6 服务器。如果网络中有一台专用 DHCPv6 服务器，该如何配置使其可以动态分配 IPv6 地址参数？让我们挑战一下吧！DHCPv6 服务器配置实验拓扑如图 7-15 所示。

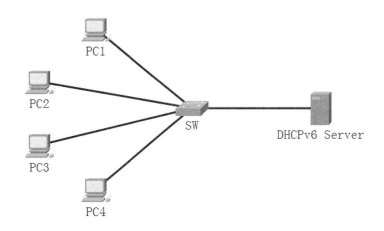

图 7-15 DHCPv6 服务器配置实验拓扑

任务要求：
- 按图 7-15 所示搭建拓扑；
- 按图 7-15 所示配置 DHCPv6 专用服务器；
- 使客户端可以动态获取 IPv6 地址并实现互通。

请读者根据任务要求独立完成挑战。

7.5 本章小结

本章内容到此结束。本章主要内容包括 IPv6 地址类型、各种类型 IPv6 地址配置以及无状态地址自动配置（SLAAC）与 DHCPv6 服务配置。IPv6 技术的诞生彻底解决了 IPv4 地址紧缺问题。在本章的学习中，我们了解了 IPv6 庞大的地址空间及各种类型 IPv6 地址的应用。本章通过 7 个应用场景和 2 个挑战练习，帮助读者熟练掌握 IPv6 地址的配置方法，做到活学活用，为后续高级 IPv6 的学习奠定基础。

第8章

学习高级 IPv6

本章要点：

- 学习 IPv6 静态路由
- 学习 IPv6 动态路由
- 学习 IPv6 ACL
- 配置 IPv6 综合案例
- 挑战练习
- 本章小结

第 7 章我们学习了 IPv6 地址等基础知识，本章我们将带领大家深入学习 IPv6 技术。其中，8.1 节介绍 IPv6 静态路由，包括 IPv6 静态路由分类及配置；8.2 节介绍 IPv6 动态路由协议 RIPng 和 OSPFv3；8.3 节介绍 IPv6 ACL 及其配置；8.4 节介绍 IPv6 综合案例，本案例综合应用了 IPv6 静态路由和 IPv6 动态路由协议 RIPng 和 OSPFv3 以及路由重发布相关知识。此外，本章通过设计的 4 个应用场景、1 个综合案例和 2 个挑战练习，使读者能够熟练应用 IPv6 技术规划和部署网络。

8.1 学习 IPv6 静态路由

8.1.1 认识 IPv6 静态路由

伴随着信息产业技术的变革，IP 由 IPv4 过渡到 IPv6。在此期间，IP 地址、路由协议以及路由表等都跟着发生了相应的变化，但是从本质上来说，IPv4 技术与 IPv6 技术没有太大的区别。因此，本节将不再对 IPv6 静态路由进行详细介绍，需要了解静态路由方面知识的读者，请阅读本书的姊妹篇《Packet Tracer 经典案例之路由交换入门篇》。

8.1.2 IPv6 静态路由分类

IPv6 静态路由的分类与《Packet Tracer 经典案例之路由交换入门篇》介绍的 IPv4 静态路由大致相同，在此简单总结如下。

- 标准静态路由：普通的、常规的通往目的网络的路由；
- 默认静态路由：将 ::/0 作为目的网络地址的路由，可匹配所有数据包，路由优先级最低；
- 汇总静态路由：将多条地址连续的网络汇总成一条路由，可减少路由条目，优化路由表；
- 浮动静态路由：提供备份路由，通过设置不同的管理距离，决定主备路由；
- 主机静态路由：优先级最高的路由，按照路由表的最长匹配原则精准匹配 128 位；
- 等价静态路由：提供备份路由，路由的管理距离相同，两条路由同时生效。

8.1.3 IPv6 静态路由语法

IPv6 静态路由配置语法：

Router(config)# **ipv6 route** *ipv6-prefix/prefix-length {ipv6-address | exit-intf | exit-intf ipv6-address}{AD}*

命令参数解释如下：

- *ipv6-prefix*：IPv6 前缀，即 IPv6 目标网络地址；
- *prefix-length*：前缀长度，取值范围为 0～128；
- *ipv6-address*：下一跳路由器的 IPv6 地址，通往目的地址的"邻居"地址；

- *exit-intf*：本地送出接口，是通往目的地址的本地送出接口，一般应用于点对点串行链路；
- *exit-intf ipv6-address*：同时指定本地送出接口和下一跳路由器的 IPv6 地址；
- *AD*：管理距离（Administrative Distance），取值范围为 1～254，255 表示不可信路由。

8.1.4 学习 IPv6 静态路由配置方法

IPv6 静态路由的配置应用举例：

（1）IPv6 直连静态路由

> R1(config)#**ipv6 route 3198:1118:930:A:: /64 Serial0/0/0**

注释：必须保证送出接口是可用的且送出接口属于点对点类型网络。

（2）IPv6 下一跳静态路由

> R1(config)#**ipv6 route 3198:1118:930:A:: /64 2198:1118:930:A::2**

注释：必须保证下一跳 IPv6 地址有效，不能使用 EUI-64 地址和链路本地址。

（3）IPv6 完全指定静态路由

> R1(config)#**ipv6 route 3198:1118:930:A:: /64 GigabitEthernet0/2 2198:1118:930:A::2**

注释：必须同时指定送出接口和下一跳 IPv6 地址，一般应用于多路访问类型网络。

8.1.5 场景一：配置 IPv6 静态路由

通过上述内容的学习，相信读者已经初步掌握了 IPv6 静态路由的相关知识，接下来，让我们一起通过具体实验来完成 IPv6 静态路由配置。配置 IPv6 静态路由实验拓扑如图 8-1 所示。

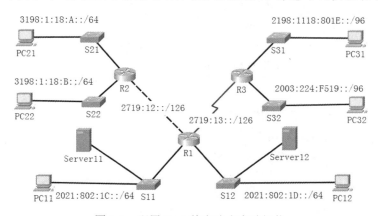

图 8-1 配置 IPv6 静态路由实验拓扑

任务要求：

- 按图 8-1 所示搭建拓扑，路由器 R1 和 R2 通过千兆位以太网接口连接，与路由器 R3 通过串行接口连接；
- 按图 8-1 所示为路由器配置 IPv6 地址，连接交换机的千兆位以太网接口采用本网最小 IPv6 地址；
- 配置 IPv6 静态路由，请采用最优方案配置 IPv6 静态路由，确保所有终端实现互通；
- 配置终端 PC IPv6 地址为 X::1，服务器地址配置为 X::254，X 代表所属的网络前缀。

任务实施如下所述。

步骤一：配置路由器接口 IPv6 地址

（1）在路由器 R1 接口上配置 IPv6 地址

```
R1(config)#ipv6 unicast-routing
R1(config)#interface GigabitEthernet0/0
R1(config-if)#ipv6 address FE80::1 link-local      //指定链路本地地址，便于识别路由器，可省略
R1(config-if)#ipv6 address 2021:802:1C::/64
R1(config-if)#no shutdown
R1(config-if)#interface GigabitEthernet0/1
R1(config-if)#ipv6 address FE80::1 link-local
R1(config-if)#ipv6 address 2021:802:1D::/64
R1(config-if)#no shutdown
R1(config-if)#interface GigabitEthernet0/2
R1(config-if)#ipv6 address FE80::1 link-local
R1(config-if)#ipv6 address 2719:12::1/126
R1(config-if)#no shutdown
R1(config-if)#interface Serial0/0/0
R1(config-if)#ipv6 address FE80::1 link-local
R1(config-if)#ipv6 address 2719:13::1/126
R1(config-if)#no shutdown
```

（2）在路由器 R2 接口上配置 IPv6 地址

```
R2(config)#ipv6 unicast-routing
R2(config)#interface GigabitEthernet0/0
R2(config-if)#ipv6 address FE80::2 link-local
R2(config-if)#ipv6 address 3198:1:18:A::/64
R2(config-if)#no shutdown
R2(config-if)#interface GigabitEthernet0/1
R2(config-if)#ipv6 address FE80::2 link-local
```

```
R2(config-if)#ipv6 address 3198:1:18:B::/64
R2(config-if)#no shutdown
R2(config-if)#interface GigabitEthernet0/2
R2(config-if)#ipv6 address FE80::2 link-local
R2(config-if)#ipv6 address 2719:12::2/126
R2(config-if)#no shutdown
```

（3）在路由器 R3 接口上配置 IPv6 地址

```
R3(config)#ipv6 unicast-routing
R3(config)#interface GigabitEthernet0/0
R3(config-if)#ipv6 address FE80::3 link-local
R3(config-if)#ipv6 address 2198:1118:801E::/96
R3(config-if)#no shutdown
R3(config-if)#interface GigabitEthernet0/1
R3(config-if)#ipv6 address FE80::3 link-local
R3(config-if)#ipv6 address 2003:224:F519::/96
R3(config-if)#no shutdown
R3(config-if)#interface Serial0/0/0
R3(config-if)#ipv6 address FE80::3 link-local
R3(config-if)#ipv6 address 2719:13::3/126
R3(config-if)#no shutdown
```

步骤二：配置路由器 IPv6 静态路由

（1）在路由器 R1 上配置 IPv6 静态路由

```
R1(config)#ipv6 route 3198:1:18:A::/63 GigabitEthernet0/2 FE80::2    //完全指定、汇总静态路由
R1(config)#ipv6 route 2198:1118:801E::/96 Serial0/0/0                //指定送出接口
R1(config)#ipv6 route 2003:224:F519::/96 Serial0/0/0                 //指定送出接口
```

（2）在路由器 R2 上配置 IPv6 静态路由

```
R2(config)#ipv6 route ::/0 2719:12::1         //默认静态路由，指定下一跳
```

（3）在路由器 R3 上配置 IPv6 静态路由

```
R3(config)#ipv6 route ::/0 Serial0/0/0        //默认静态路由，指定送出接口
```

步骤三：查看路由器的路由表信息

（1）在路由器 R1 上查看 IPv6 路由表

```
R1>show ipv6 route
```

```
IPv6 Routing Table - 12 entries
<output omitted>
S    2003:224:F519::/96 [1/0]
        via Serial0/0/0, directly connected        //IPv6 直连静态路由，送出接口是串行接口
C    2021:802:1C::/64 [0/0]
        via GigabitEthernet0/0, directly connected
L    2021:802:1C::/128 [0/0]
        via GigabitEthernet0/0, receive
C    2021:802:1D::/64 [0/0]
        via GigabitEthernet0/1, directly connected
L    2021:802:1D::/128 [0/0]
        via GigabitEthernet0/1, receive
S    2198:1118:801E::/96 [1/0]
        via Serial0/0/0, directly connected        //IPv6 直连静态路由，送出接口是串行接口
C    2719:12::/126 [0/0]
        via GigabitEthernet0/2, directly connected
L    2719:12::1/128 [0/0]
        via GigabitEthernet0/2, receive
C    2719:13::/126 [0/0]
        via Serial0/0/0, directly connected
L    2719:13::1/128 [0/0]
        via Serial0/0/0, receive
S    3198:1:18:A::/63 [1/0]
        via FE80::2, GigabitEthernet0/2            //IPv6 完全指定静态路由，IPv6 汇总静态路由
L    FF00::/8 [0/0]
        via Null0, receive
```

（2）在路由器 R2 上查看 IPv6 路由表

```
R2>show ipv6 route
IPv6 Routing Table - 8 entries
<output omitted>
S    ::/0 [1/0]
        via 2719:12::1          // IPv6 默认静态路由，IPv6 下一跳静态路由
<output omitted>
```

（3）在路由器 R3 上查看 IPv6 路由表

```
R3>show ipv6 route
IPv6 Routing Table - 8 entries
```

```
<output omitted>
S    ::/0 [1/0]
        via Serial0/0/0, directly connected    // IPv6 默认静态路由，IPv6 直连静态路由
<output omitted>
```

步骤四：测试网络连通性

此时终端可互访，请读者自行测试。以下展示了主机 PC11 跟踪 PC31 的数据包跟踪路径。

```
C:\>tracert 2198:1118:801E::1
Tracing route to 2198:1118:801E::1 over a maximum of 30 hops:
    1    0ms    0ms    0ms    2021:802:1C::          //路由器 R1
    2    5ms    11ms   10ms   2719:13::3             //路由器 R3
    3    0ms    0ms    11ms   2198:1118:801E::1      //主机 PC31
Trace complete.
```

以下展示了主机 PC21 跟踪服务器 Server12 的数据包跟踪路径。

```
C:\>tracert 2021:802:1D::254
Tracing route to 2021:802:1D::254 over a maximum of 30 hops:
    1    0ms    0ms    0ms    3198:1:18:A::          //路由器 R2
    2    0ms    0ms    0ms    2719:12::1             //路由器 R1
    3    0ms    0ms    0ms    2021:802:1D::254       //服务器 Server12
Trace complete.
```

至此，IPv6 静态路由配置实验已经顺利完成。请读者在路由器 R2 和 R3 上再添加一条串行链路，请思考此时该如何配置和修改 IPv6 静态路由。请读者独立完成。

8.2 学习 IPv6 动态路由

8.2.1 认识 RIPng 路由协议

RIPng（RIP next generation）是 RIP 的 IPv6 版本。RIPng 和 RIPv2 类似，有相同的算法、计时器、消息类型，以及相同方式请求和响应消息等。RIPng 的特点总结如下：
- RIPng 基于 UDP，采用 521 端口号收发数据报；
- RIPng 是距离矢量路由协议，默认管理距离为 120；
- RIPng 度量为跳数，最大跳数为 15，16 跳的网络不可达；
- RIPng 使用链路本地地址 FE80::/10 作为源地址发送路由更新信息；
- RIPng 使用 FF02::9 组播地址更新信息，每 30 s 向邻居发送完整路由表；

- RIPng 没有认证机制，靠 IPv6 安全机制（扩展报头）保证报文合法。

8.2.2 学习 RIPng 配置语法

RIPng 主要配置语法：

Router(config)#**ipv6 router rip** *name*	//创建 RIPng 进程
Router(config-rtr)#**interface** *interface*	
Router(config-if)#**ipv6 rip** *name* **enable**	//接口启用 RIP 进程

8.2.3 场景二：配置 RIPng 动态路由

通过 8.2.1 节和 8.2.2 节内容的学习，相信大家已经对 RIPng 路由协议有了初步认知，接下来，让我们一起通过具体实验来完成 RIPng 配置。配置 RIPng 实验拓扑如图 8-2 所示。

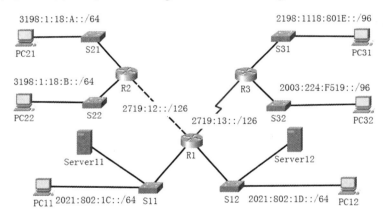

图 8-2　配置 RIPng 实验拓扑

任务要求：

- 按图 8-2 所示搭建拓扑，路由器 R1 和 R2 通过千兆位以太网接口连接，与路由器 R3 通过串行接口连接；
- 按图 8-2 所示为路由器配置 IPv6 地址，连接交换机的千兆位以太网接口采用本网最小 IPv6 地址；
- 配置终端 PC IPv6 地址为 X::1，服务器地址配置为 X::254，X 代表所属的网络前缀；
- 配置动态路由协议 RIPng，实现全网互通。

任务实施如下所述。

步骤一：配置路由器接口 IPv6 地址

3 台路由器接口 IPv6 地址的配置与 8.1.5 节场景一中的步骤一完全相同。

步骤二：配置 RIPng 动态路由协议

（1）在路由器 R1 上配置 RIPng

```
R1(config)#ipv6 router rip 1           ///创建 RIPng 进程 1
R1(config-rtr)#interface GigabitEthernet 0/0
R1(config-if)#ipv6 rip 1 enable        //启用 RIPng 进程 1
R1(config-if)#interface GigabitEthernet 0/1
R1(config-if)#ipv6 rip 1 enable
R1(config-if)#interface GigabitEthernet 0/2
R1(config-if)#ipv6 rip 1 enable
R1(config-if)#interface Serial 0/0/0
R1(config-if)#ipv6 rip 1 enable
```

（2）在路由器 R2 上配置 RIPng

```
R2(config)#ipv6 router rip 1
R2(config-rtr)#interface GigabitEthernet 0/0
R2(config-if)#ipv6 rip 1 enable
R2(config-if)#interface GigabitEthernet 0/1
R2(config-if)#ipv6 rip 1 enable
R2(config-if)#interface GigabitEthernet 0/2
R2(config-if)#ipv6 rip 1 enable
```

（3）在路由器 R3 上配置 RIPng

```
R3(config)#ipv6 router rip 1
R3(config-rtr)#interface GigabitEthernet 0/0
R3(config-if)#ipv6 rip 1 enable
R3(config-if)#interface GigabitEthernet 0/1
R3(config-if)#ipv6 rip 1 enable
R3(config-if)#interface Serial 0/0/0
R3(config-if)#ipv6 rip 1 enable
```

步骤三：查看路由器的路由表信息

（1）在路由器 R1 上查看 IPv6 路由表

```
R1>show ipv6 route
IPv6 Routing Table - 13 entries
<output omitted>
R    2003:224:F519::/96 [120/2]
```

```
                via FE80::3, Serial0/0/0            //下一跳为R3，到达目标网络度量为2跳
    <output omitted>
    R   2198:1118:801E::/96 [120/2]
                via FE80::3, Serial0/0/0            //下一跳为R3，到达目标网络度量为2跳
    R   3198:1:18:A::/64 [120/2]
                via FE80::2, GigabitEthernet0/2     //下一跳为R2，到达目标网络度量为2跳
    R   3198:1:18:B::/64 [120/2]
                via FE80::2, GigabitEthernet0/2     //下一跳为R2，到达目标网络度量为2跳
    <output omitted>
```

以上 R1 的 IPv6 路由表输出表明，R1 添加了 4 条从邻居 R3 和 R2 学习到的 RIP 路由。

（2）在路由器 R2 上查看 IPv6 路由表

```
R2>show ipv6 route
IPv6 Routing Table - 12 entries
<output omitted>
    R   2003:224:F519::/96 [120/3]
                via FE80::1, GigabitEthernet0/2     //下一跳为R1，到达目标网络度量为3跳
    R   2021:802:1C::/64 [120/2]
                via FE80::1, GigabitEthernet0/2     //下一跳为R1，到达目标网络度量为2跳
    R   2021:802:1D::/64 [120/2]
                via FE80::1, GigabitEthernet0/2     //下一跳为R1，到达目标网络度量为2跳
    R   2198:1118:801E::/96 [120/3]
                via FE80::1, GigabitEthernet0/2     //下一跳为R1，到达目标网络度量为3跳
    <output omitted>
    R   2719:13::/126 [120/2]
                via FE80::1, GigabitEthernet0/2     //下一跳为R1，到达目标网络度量为2跳
    <output omitted>
```

以上 R2 的 IPv6 路由表输出表明，R2 添加了 5 条从邻居 R1 学习到的 RIP 路由。

（3）在路由器 R3 上查看 IPv6 路由表

```
R3>show ipv6 route summary
IPv6 routing table name is default(0) global scope - 12 entries
IPv6 routing table default maximum-paths is 16
Route Source        Networks        Overhead        Memory (bytes)
connected           3               264             372
local               4               352             496
rip                 5               440             620
Total               12              1056            1488
    Number of prefixes:
```

/8: 1, /64: 4, /96: 2, /126: 2, /128: 3

以上 IPv6 路由汇总信息表明，R3 学到 5 条 RIP 路由。

步骤四：测试网络连通性

此时终端可以互访，请读者自行测试。以下展示了主机 PC12 跟踪 PC21 的数据包跟踪路径。

```
C:\>tracert 3198:1:18:A::1
Tracing route to 3198:1:18:A::1 over a maximum of 30 hops:
  1    0ms    0ms    0ms    2021:802:1D::         //路由器 R1
  2    0ms    0ms    0ms    2719:12::2            //路由器 R2
  3    0ms    0ms    0ms    3198:1:18:A::1        //主机 PC21
Trace complete.
```

以下展示了主机 PC22 跟踪主机 PC32 的数据包跟踪路径。

```
C:\>tracert 2003:224:F519::1
Tracing route to 2003:224:F519::1 over a maximum of 30 hops:
  1    0ms    5ms    0ms    3198:1:18:B::         //路由器 R2
  2    0ms    0ms    5ms    2719:12::1            //路由器 R1
  3    0ms    6ms    5ms    2719:13::3            //路由器 R3
  4    1ms    5ms    1ms    2003:224:F519::1      //主机 PC32
Trace complete.
```

至此，动态路由协议 RIPng 配置实验结束。RIPng 配置相比 RIP 有较大改变，取消了 IPv4 的 **network** 命令，改成在接口上调用 RIP 进程，从而简化了配置。

8.2.4 认识 OSPFv3 路由协议

OSPFv3 在 RFC 2740 中发布并有详细描述。OSPFv3 与 OSPFv2 的关系非常类似于 RIPng 与 RIPv2 的关系。OSPFv3 通过 IPv6 网络层运行，通告 IPv6 前缀；OSPFv2 运行在 IPv4 网络中，通告 IPv4 路由。OSPFv3 使用与 OSPFv2 相同的基本实现机制，包括 SPF 算法、消息类型、DR 选举、分层区域设计以及度量标准等。OSPFv3 与 OSPFv2 重要不同之处在于以下几个方面：

- OSPFv3 将 OSPFv2 "子网"概念改变为"链路"，一条链路的接口可以拥有多个 IPv6 地址，单条链路可以属于多个 IPv6 子网，即属于同一链路但属于不同 IPv6 子网的两个邻居仍然可以交换数据包并进行通信。
- OSPFv3 LSA 有路由器 LSA、网络 LSA、区域间前缀 LSA、区域间路由器 LSA、AS 外部 LSA、组成员 LSA 和类型 7 LSA，新增链路 LSA 和区域内前缀 LSA。路由器 LSA 和网络 LSA 取消通告前缀功能，这两种 LSA 对 SPF 算法来说只代表路由器节点信息，通告前缀功能被放入新增的区域内前缀 LSA 中。OSPFv3 将邻居特有信息放入新的链路 LSA 中，链路 LSA 只有链路本地扩散范围。

- OSPFv3 报头取消认证字段，本身不支持认证，认证通过 IPv6 报文扩展报头实现。OSPFv3 的 Hello 和 DBD 数据包格式与 OSPFv2 不同。OSPFv3 Hello 数据包取消网络掩码字段，其报文无效时间间隔从原来 32 位减少到 16 位。Hello 和 DBD 数据包的可选项字段由原来 8 位增大到 24 位。
- OSPFv3 路由器 ID、区域 ID 依然保留在 IPv6 中，沿用 OSPFv2 的 32 位，采用点分十进制表示。OSPFv3 邻居通过路由器 ID 标识。
- OSPFv3 报文头部增加实例 ID 字段，实例 ID 用于识别 OSPFv3 的不同实例。OSPFv3 允许每条链路运行多个 OSPFv3 实例。
- OSPFv3 采用 FF02::5 和 FF02::6 作为组播地址，OSPFv2 采用 224.0.0.5 和 224.0.0.6 作为组播地址。
- OSPFv3 使用路由器链路本地 IPv6 单播地址（FE80::/10 开头）作为源地址和下一跳。

8.2.5　学习 OSPFv3 配置语法

OSPFv3 路由主要配置语法如下：

```
Router(config)#ipv6 router ospf process              //创建 OSPFv3 进程
Router(config-rtr)#router-id ipv4-address            //指定路由器 ID，路由器 ID 采用 IPv4 地址格式
Router(config-rtr)#interface interface
Router(config-if)#ipv6 ospf process area area        //接口启用 OSPFv3 进程并加入指定区域
```

8.2.6　场景三：配置 OSPFv3 动态路由

通过 8.2.1 节和 8.2.2 节内容的学习，相信大家对 OSPFv3 已经有了初步认知，接下来，让我们一起通过具体实验来完成 OSPFv3 配置。配置 OSPFv3 路由实验拓扑如图 8-3 所示。

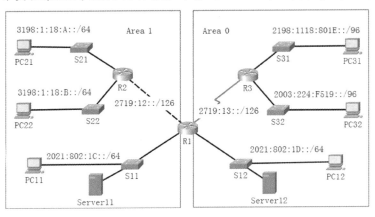

图 8-3　配置 OSPFv3 路由实验拓扑

任务要求：
- 按图 8-3 所示搭建拓扑，路由器 R1 和 R2 通过千兆位以太网接口连接，与路由器 R3 通过串行接口连接；
- 按图 8-3 所示为路由器配置 IPv6 地址，连接交换机的千兆位以太网接口采用本网最小 IPv6 地址；
- 配置终端 PC IPv6 地址为 X::1，服务器地址配置为 X::254，X 代表所属的网络前缀；
- 配置动态路由协议 OSPFv3，实现全网互通。

步骤一：配置路由器接口 IPv6 地址

3 台路由器接口 IPv6 地址的配置与 8.1.5 节场景一中的步骤一完全相同。

步骤二：配置动态路由协议 OSPFv3

（1）在路由器 R1 上配置 OSPFv3

```
R1(config)#ipv6 router ospf 1              //创建 OSPFv3 进程 1
R1(config-rtr)#router-id 1.1.1.1           //指定路由器 ID
R1(config-rtr)#interface GigabitEthernet 0/0
R1(config-if)#ipv6 ospf 1 area 1           //启用 OSPFv3 进程 1，将接口加入区域 1
R1(config-if)#interface GigabitEthernet 0/2
R1(config-if)#ipv6 ospf 1 area 1
R1(config-if)#interface GigabitEthernet 0/1
R1(config-if)#ipv6 ospf 1 area 0
R1(config-if)#interface Serial 0/0/0
R1(config-if)#ipv6 ospf 1 area 0
```

（2）在路由器 R2 上配置 OSPFv3

```
R2(config)#ipv6 router ospf 1
R2(config-rtr)#router-id 2.2.2.2
R2(config-rtr)#interface GigabitEthernet 0/0
R2(config-if)#ipv6 ospf 1 area 1
R2(config-if)#interface GigabitEthernet 0/1
R2(config-if)#ipv6 ospf 1 area 1
R2(config-if)#interface GigabitEthernet 0/2
R2(config-if)#ipv6 ospf 1 area 1
```

（3）在路由器 R3 上配置 OSPFv3

```
R3(config)#ipv6 router ospf 1
```

```
R3(config-rtr)#router-id 3.3.3.3
R3(config-rtr)#interface GigabitEthernet 0/0
R3(config-if)#ipv6 ospf 1 area 0
R3(config-if)#interface GigabitEthernet 0/1
R3(config-if)#ipv6 ospf 1 area 0
R3(config-if)#interface Serial 0/0/0
R3(config-if)#ipv6 ospf 1 area 0
```

步骤三：查看 OSPFv3 的三张表

（1）查看 OSPFv3 邻居表

```
R1>show ipv6 ospf neighbor
```

Neighbor ID	Pri	State	Dead Time	Interface ID	Interface
2.2.2.2	1	FULL/BDR	00:00:33	3	GigabitEthernet0/2
3.3.3.3	0	FULL/ -	00:00:31	4	Serial0/0/0

（2）查看 OSPFv3 拓扑表

```
R2>show ipv6 ospf database
```

OSPF Router with ID (2.2.2.2) (Process ID 1)

Router Link States (Area 1)　　//路由器 LSA

ADV Router	Age	Seq#	Fragment ID	Link count	Bits
2.2.2.2	1012	0x80000002	0	1	
1.1.1.1	1012	0x80000002	0	1	B

Net Link States (Area 1)　　//网络 LSA

ADV Router	Age	Seq#	Link ID (DR)	Rtr count
1.1.1.1	1012	0x80000001	3	2

Inter Area Prefix Link States (Area 1)　　//区域内前缀 LSA

ADV Router	Age	Seq#	Metric	Prefix
1.1.1.1	1057	0x80000001	1	2021:802:1D::/64
1.1.1.1	1057	0x80000002	64	2719:13::/126
1.1.1.1	962	0x80000003	65	2198:1118:801E::/96
1.1.1.1	962	0x80000004	65	2003:224:F519::/96

Link (Type-8) Link States (Area 1)　　//链路 LSA

ADV Router	Age	Seq#	Link ID	Interface
2.2.2.2	1013	0x80000001	1	Gi0/0
2.2.2.2	1013	0x80000002	2	Gi0/1
2.2.2.2	1012	0x80000004	3	Gi0/2
1.1.1.1	1013	0x80000003	3	Gi0/2

Intra Area Prefix Link States (Area 1)　　//区域间前缀 LSA

ADV Router	Age	Seq#	Link ID	Ref-lstype	Ref-LSID
1.1.1.1	1012	0x80000003	1	0x2002	3
1.1.1.1	1012	0x80000004	2	0x2001	0
2.2.2.2	974	0x80000004	2	0x2001	0

（3）查看 OSPFv3 路由表

```
R1>show ipv6 route ospf
<output omitted>
O    2003:224:F519::/96 [110/65]
     via FE80::3, Serial0/0/0            //域内路由，下一跳为R3，到达目标网络开销为65
O    2198:1118:801E::/96 [110/65]
     via FE80::3, Serial0/0/0            //域内路由，下一跳为R3，到达目标网络开销为65
O    3198:1:18:A::/64 [110/2]
     via FE80::2, GigabitEthernet0/2     //域内路由，下一跳为R2，到达目标网络开销为2
O    3198:1:18:B::/64 [110/2]
     via FE80::2, GigabitEthernet0/2     //域内路由，下一跳为R2，到达目标网络开销为2
```

以上输出表明，路由器 R1 的 IPv6 路由表中添加了 4 条从邻居路由器 R2 和 R3 学习到的 OSPF 域内路由。

```
R2>show ipv6 route ospf
IPv6 Routing Table - 12 entries
<output omitted>
OI   2003:224:F519::/96 [110/66]
     via FE80::1, GigabitEthernet0/2     //域间路由，下一跳为R1，到达目标网络开销为66
O    2021:802:1C::/64 [110/2]
     via FE80::1, GigabitEthernet0/2     //域内路由，下一跳为R1，到达目标网络开销为2
OI   2021:802:1D::/64 [110/2]
     via FE80::1, GigabitEthernet0/2     //域间路由，下一跳为R1，到达目标网络开销为2
OI   2198:1118:801E::/96 [110/66]
     via FE80::1, GigabitEthernet0/2     //域间路由，下一跳为R1，到达目标网络开销为66
OI   2719:13::/126 [110/65]
     via FE80::1, GigabitEthernet0/2     //域间路由，下一跳为R1，到达目标网络开销为65
```

以上输出表明，路由器 R2 的 IPv6 路由表中添加了 5 条从邻居 R1 学习到的 OSPF 路由，其中，域内路由（Area 1）1 条，域间路由（Area 0）4 条。

```
R3>show ipv6 route ospf
IPv6 Routing Table - 12 entries
```

```
<output omitted>
OI   2021:802:1C::/64 [110/65]
     via FE80::1, Serial0/0/0      //域间路由，下一跳为 R1，到达目标网络开销为 65
O    2021:802:1D::/64 [110/65]
     via FE80::1, Serial0/0/0      //域内路由，下一跳为 R1，到达目标网络开销为 65
OI   2719:12::/126 [110/65]
     via FE80::1, Serial0/0/0      //域间路由，下一跳为 R1，到达目标网络开销为 65
OI   3198:1:18:A::/64 [110/66]
     via FE80::1, Serial0/0/0      //域间路由，下一跳为 R1，到达目标网络开销为 66
OI   3198:1:18:B::/64 [110/66]
     via FE80::1, Serial0/0/0      //域间路由，下一跳为 R1，到达目标网络开销为 66
```

以上输出表明，路由器 R3 的 IPv6 路由表中添加了 5 条从邻居 R1 学习到的 OSPF 路由，其中，域内路由（Area 0）1 条，域间路由（Area 1）4 条。

步骤四：测试网络连通性

此时终端可互访，请读者自行测试。以下展示了路由器 R2 跟踪 Server12 的数据包跟踪路径。

```
R2>traceroute 2021:802:1D::254
Tracing the route to 2021:802:1D::254
   1   2719:12::1         0msec   0msec   0msec    //路由器 R1
   2   2021:802:1D::254   0msec   0msec   0msec    //服务器 Server12
Minimum = 2ms, Maximum = 12ms, Average = 6ms
```

以下展示了路由器 R2 跟踪主机 PC31 的数据包跟踪路径。

```
R2>traceroute 2198:1118:801E::1
Tracing the route to 2198:1118:801E::1
   1   2719:12::1           0msec   0msec   0msec    //路由器 R1
   2   2719:13::3           5msec   0msec   7msec    //路由器 R3
   3   2198:1118:801E::1    1msec   5msec   1msec    //主机 PC31
```

至此，动态路由协议 OSPFv3 配置实验结束。OSPFv3 配置相比 OSPFv2 有较大改变，取消 IPv4 的 **network** 命令，改成在接口上调用 OSPF 进程，从而简化了配置。但是，OSPFv3 依然采用 IPv4 地址格式的路由器 ID，且必须手动指定。

8.3 学习 IPv6 ACL

在第 5 章我们学习了 IPv4 ACL 的技术及其应用，本节我们将学习 IPv6 ACL。

8.3.1 认识 IPv6 ACL

IPv6 ACL 在操作和配置方面类似 IPv4 ACL，熟悉 IPv4 ACL 会很容易理解和配置 IPv6 ACL。尽管 IPv4 ACL 和 IPv6 ACL 相似，但它们之间还是有区别的，主要区别如下：

① IPv4 ACL 分标准 ACL 和扩展 ACL 两种类型，且可以是编号或命名 ACL；IPv6 ACL 只有一种类型，只能采用命名 ACL，等同于 IPv4 扩展命名 ACL。

② IPv4 ACL 结尾有一条隐式语句 deny any 或 deny ip any any；IPv6 ACL 结尾隐式语句是 deny ip any any，还包含两条隐式语句 permit icmp any any nd-na 和 permit icmp any any nd-ns 语句，即同时允许"邻居发现-邻居通告"和"邻居发现-邻居请求"消息。

③ IPv4 ACL 使用命令 ip access-group {number | name} {in | out} 将 IPv4 ACL 应用于接口；IPv6 ACL 使用命令 ipv6 traffic-filter name {in | out} 将 IPv6 ACL 应用于接口。

注意：IPv4 和 IPv6 ACL 都使用 **access-class** 命令将 ACL 应用于 VTY 线路。

8.3.2 学习 IPv6 ACL 语法

IPv6 ACL 主要配置语法如下：

```
Router(config)#ipv6 access-list name                     //创建 IPv6 ACL
Router(config-ipv6-acl)#deny | permit protocol {sourcce-ipv6-prefix/prefix-length | any | host source-ipv6-address }[operator [port-number]] {destination-ipv6-prefix/prefix-length | any | host destination-ipv6-address }[operator [port-number]]          //创建匹配规则
Router(config-ipv6-acl)#remark remark                    //注释 IPv6 ACL
Router(config-ipv6-acl)#interface interface
Router(config-if)#ipv6 traffic-filter name { in | out }  //应用 IPv6 ACL
```

- *protocol*：协议的名称或编号，如 ipv6、icmp、tcp、udp 等；
- *sourcce-ipv6-prefix/prefix-length*：源 IPv6 前缀 / 前缀长度；
- *destination-ipv6-prefix/prefix-length*：目的 IPv6 前缀 / 前缀长度；
- *any*：匹配所有地址，代表 IPv6 地址::/0；
- *host*：后面跟源主机或目的主机 IPv6 地址；
- *operator*：运算符，包括 lt（小于）、gt（大于）、eq（等于）、neq（不等于）等；
- *port-number*：端口号，用于过滤 TCP 或 UDP 的具体应用端口或名称。

8.3.3 场景四：配置 IPv6 ACL

通过 8.3.1 节和 8.3.2 节内容的学习，相信大家都已经对 IPv6 ACL 有了初步认知，接下来，让我们一起通过具体实验来完成 IPv6 ACL 配置。配置 IPv6 ACL 实验拓扑如图 8-4 所示。

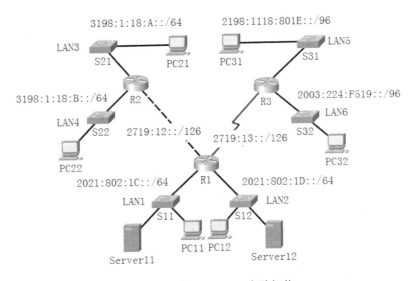

图 8-4 配置 IPv6 ACL 实验拓扑

任务要求：
- 按图 8-4 所示搭建拓扑，路由器 R1 和 R2 通过千兆位以太网接口连接，与路由器 R3 通过串行口连接；
- 按图 8-4 配置路由器 IPv6 地址，连接交换机的千兆位以太网接口采用本网内最小 IPv6 地址；
- 配置终端 PC IPv6 地址为 X::1，将服务器地址配置为 X::254，X 代表所属网络前缀；
- 配置动态路由协议 RIPng 或 OSPFv3，要求实现全网互通；
- 配置 IPv6 ACL，要求 Server11 和 Server12 禁止 LAN3、LAN4 和 LAN5 的主机发送的 ping 包，其他访问不受限制；LAN5 仅允许 LAN6，以及主机 PC12 和 PC21 访问；
- 请按 IPv6 ACL 限制要求，依次进行访问效果测试。

任务实施如下所述。

步骤一：配置路由器 IPv6 地址

3 台路由器接口 IPv6 地址配置与 8.1.5 节场景一中的步骤一完全相同。

步骤二：配置动态路由协议

请选择动态路由协议 RIPng 或 OSPFv3 配置网络设备，确保实现全网互通，终端可以互访。

步骤三：配置 IPv6 ACL 限制访问服务器

（1）终端到服务器的连通性测试

请确保所有 PC 均可以与服务器 Server11 和 Server12 通信。请读者分别采用命令行和浏览器模式进行连通性测试。以下输出表明,LAN3～LAN6 的主机与服务器 Server11 的 ping 测试成功。

```
C:\>ping 2021:802:1C::254

Pinging 2021:802:1C::254 with 32 bytes of data:
Reply from 2021:802:1C::254: bytes=32 time<1ms TTL=126
Reply from 2021:802:1C::254: bytes=32 time<1ms TTL=126
Reply from 2021:802:1C::254: bytes=32 time<1ms TTL=126
Reply from 2021:802:1C::254: bytes=32 time<1ms TTL=126

Ping statistics for 2021:802:1C::254:
    Packets: Sent = 4, Received = 4, Lost = 0 (0% loss),
Approximate round trip times in milli-seconds:
    Minimum = 0ms, Maximum = 0ms, Average = 0ms
```

(2) 在路由器 R1 上配置 IPv6 ACL

```
R1(config)#ipv6 access-list NO-PING-SERVER-A              //创建 IPv6 ACL
R1(config-ipv6-acl)#remark NOPING LAN3-LAN4 TO SERVER     //注释 IPv6 ACL
R1(config-ipv6-acl)#deny icmp 3198:1:18:A::/63 host 2021:802:1C::254
R1(config-ipv6-acl)#deny icmp 3198:1:18:A::/63 host 2021:802:1D::254
R1(config-ipv6-acl)#permit ipv6 any any
R1(config-ipv6-acl)#ipv6 access-list NO-PING-SERVER-B
R1(config-ipv6-acl)#remark NOPING LAN5 TO SERVER
R1(config-ipv6-acl)#deny icmp 2198:1118:801E::/96 host 2021:802:1C::254
R1(config-ipv6-acl)#deny icmp 2198:1118:801E::/96 host 2021:802:1D::254
R1(config-ipv6-acl)#permit ipv6 any any
R1(config-ipv6-acl)#interface GigabitEthernet 0/2
R1(config-if)#ipv6 traffic-filter NO-PING-SERVER-A in     //在接口上应用 IPv6 ACL
R1(config-if)#interface Serial0/0/0
R1(config-if)#ipv6 traffic-filter NO-PING-SERVER-B in
```

(3) 在路由器 R1 上查看 IPv6 ACL

```
R1#show ipv6 access-list
IPv6 access list NO-PING-SERVER-A
    deny icmp 3198:1:18:A::/63 host 2021:802:1C::254
    deny icmp 3198:1:18:A::/63 host 2021:802:1D::254
```

            permit ipv6 any any
    IPv6 access list **NO-PING-SERVER-B**
            deny icmp 2198:1118:801E::/96 host 2021:802:1C::254
            deny icmp 2198:1118:801E::/96 host 2021:802:1D::254
            permit ipv6 any any

（4）测试 IPv6 ACL 访问控制效果

以下输出表明，LAN5 中的主机 PC31 发往服务器 Server12 的 ping 包被路由器 R1 的 ACL 拦截。请读者自行测试其余网段对服务器 Server11 和 Server12 的连通性。

    C:\>**ping 2021:802:1D::254**
    Pinging 2021:802:1D::254 with 32 bytes of data:
    Reply from 2719:13::1: Destination host unreachable.
    Reply from 2719:13::1: Destination host unreachable.
    Reply from 2719:13::1: Destination host unreachable.
    Reply from 2719:13::1: Destination host unreachable.
    Ping statistics for 2021:802:1D::254:
        Packets: Sent = 4, Received = 0, Lost = 4 (100% loss),

步骤四：配置 IPv6 ACL 限制访问网段

（1）设备到 LAN5 的连通性测试

此时，除服务器 Server11 与 Server12 外，网络中所有主机和路由器均可以访问 LAN5，请读者自行完成测试。以下输出表明，LAN1 中的主机 PC11 发出的数据包顺利到达 LAN5 主机 PC31。

    C:\>**tracert 2198:1118:801E::1**
    Tracing route to 2198:1118:801E::1 over a maximum of 30 hops:

        1   0ms   0ms   0ms   **2021:802:1C::**
        2   10ms   10ms   5ms   **2719:13::3**
        3   5ms   0ms   0ms   **2198:1118:801E::1**
    Trace complete.

以下输出表明，从路由器 R2 发出的数据包顺利到达 LAN5 中的主机 PC31。

    R2>**traceroute 2198:1118:801E::1**
    Type escape sequence to abort.
    Tracing the route to **2198:1118:801E::1**

1	2719:12::1 0msec	0msec	0msec	
2	2719:13::3 0msec	5msec	5msec	
3	2198:1118:801E::1	10msec	0msec	10msec

（2）在路由器 R3 上配置 IPv6 ACL

```
R3(config)#ipv6 access-list NO-ACCESS-LAN5
R3(config-ipv6-acl)#remark ONLY Permit LAN6-HOST_PC12_PC21 ACCESSS
R3(config-ipv6-acl)#permit ipv6 2003:224:F519::/96 2198:1118:801E::/96
R3(config-ipv6-acl)#permit ipv6 host 2021:802:1D::1 2198:1118:801E::/96
R3(config-ipv6-acl)#permit ipv6 host 3198:1:18:A::1 2198:1118:801E::/96
R3(config-ipv6-acl)#interface GigabitEthernet 0/0
R3(config-if)#ipv6 traffic-filter NO-ACCESS-LAN5 out
```

（3）测试 IPv6 ACL 访问控制效果

以下输出表明，LAN1 中的主机 PC11 发往 LAN5 中的主机 PC31 的 ping 包被路由器 R3 上配置的 ACL 拦截。

```
C:\>ping 2198:1118:801E::1
Pinging 2198:1118:801E::1 with 32 bytes of data:
Reply from 2719:13::3: Destination host unreachable.
Reply from 2719:13::3: Destination host unreachable.
Reply from 2719:13::3: Destination host unreachable.
Reply from 2719:13::3: Destination host unreachable.
Ping statistics for 2198:1118:801E::1:
    Packets: Sent = 4, Received = 0, Lost = 4 (100% loss),
```

以下输出表明，路由器 R2 发往 LAN5 中的主机 PC31 的 ping 包被路由器 R3 上配置的 ACL 拦截。

```
R2>ping 2198:1118:801E::1
Type escape sequence to abort.
Sending 5, 100-byte ICMP Echos to 2198:1118:801E::1, timeout is 2 seconds:
AAAAA
Success rate is 0 percent (0/5)
```

请读者自行完成 LAN6、主机 PC12 和 PC21 对 LAN5 的连通性测试，它们的数据包是被放行的，除此之外的主机均无法访问 LAN5。

8.4 配置 IPv6 综合案例

通过本章的学习，相信大家已经初步掌握了本章的内容，接下来，让我们一起通过具体综合实验来加深对本章知识的理解。IPv6 综合实验拓扑如图 8-5 所示。

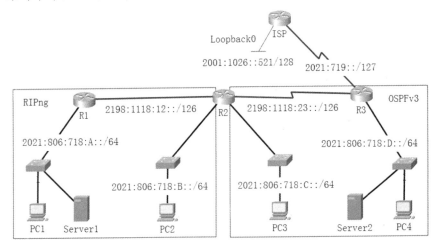

图 8-5 IPv6 综合实验拓扑

任务要求：

- 按图 8-5 所示搭建拓扑，要求路由器型号采用 4000 系列。路由器 R2 与 R1 通过千兆位光接口 GigabitEthernet0/0/2 连接，路由器 R2 与路由器 R3 通过串行接口 Serial0/2/0 连接，路由器 R3 与 ISP 通过串行接口 Serial0/2/1 连接，路由器与交换机间采用千兆位以太网接口实现连接；
- 按图 8-5 所示配置路由器 IPv6 地址，连接交换机的千兆位以太网接口采用本网内最小 IPv6 地址；
- 将终端 PC 的 IPv6 地址配置为 X::1，将服务器的地址配置为 X::254，X 代表所属网络前缀；
- 配置动态路由，要求路由器 R1 和 R2 间配置 RIPng，R2 与 R3 间配置单区域的 OSPFv3；
- 配置静态路由，要求路由器 R3 和 ISP 间配置 IPv6 静态路由，采用最优方案进行配置；
- 配置路由重分布，在合适的路由器上，将 RIP 域的路由重分布到 OSPF 域；
- 进行网络连通性测试，要求实现全网互通。

任务实施如下所述。

步骤一：配置路由器 IPv6 地址

（1）在路由器 R1 接口上配置 IPv6 地址

```
R1(config)#ipv6 unicast-routing
R1(config)#interface GigabitEthernet0/0/0
R1(config-if)#ipv6 address fe80::1 link-local
R1(config-if)#ipv6 address 2021:806:718:A::/64
R1(config-if)#no shutdown
R1(config-if)#interface GigabitEthernet0/0/2
R1(config-if)#ipv6 address fe80::1 link-local
R1(config-if)#ipv6 address 2198:1118:12::1/126
R1(config-if)#no shutdown
```

（2）在路由器 R2 接口上配置 IPv6 地址

```
R2(config)#ipv6 unicast-routing
R2(config)#interface GigabitEthernet0/0/0
R2(config-if)#ipv6 address fe80::2 link-local
R2(config-if)#ipv6 address 2021:806:718:B::/64
R2(config-if)#no shutdown
R2(config-if)#interface GigabitEthernet0/0/1
R2(config-if)#ipv6 address fe80::2 link-local
R2(config-if)#ipv6 address 2021:806:718:C::/64
R2(config-if)#no shutdown
R2(config-if)#interface GigabitEthernet0/0/2
R2(config-if)#ipv6 address fe80::2 link-local
R2(config-if)#ipv6 address 2198:1118:12::2/126
R2(config-if)#no shutdown
R2(config-if)#interface Serial0/2/0
R2(config-if)#ipv6 address fe80::2 link-local
R2(config-if)#ipv6 address 2198:1118:23::2/126
R2(config-if)#no shutdown
```

（3）在路由器 R3 接口上配置 IPv6 地址

```
R3(config)#ipv6 unicast-routing
R3(config)#interface GigabitEthernet0/0/0
R3(config-if)#ipv6 address fe80::3 link-local
R3(config-if)#ipv6 address 2021:806:718:D::/64
```

```
R3(config-if)#no shutdown
R3(config-if)#interface Serial0/2/0
R3(config-if)#ipv6 address fe80::3 link-local
R3(config-if)#ipv6 address 2198:1118:23::3/126
R3(config-if)#no shutdown
R3(config-if)#interface Serial0/2/1
R3(config-if)#ipv6 address fe80::3 link-local
R3(config-if)#ipv6 address 2021:719::/127
R3(config-if)#no shutdown
```

（4）在路由器 ISP 接口上配置 IPv6 地址

```
ISP(config)#ipv6 unicast-routing
ISP(config)#interface Serial0/2/1
ISP(config-if)#ipv6 address fe80::255 link-local
ISP(config-if)#ipv6 address 2021:719::1/127
ISP(config-if)#no shutdown
ISP(config-if)#interface Loopback 0
ISP(config-if)#ipv6 address 2001:1026::521/128
```

步骤二：配置 RIPng

（1）在路由器 R1 上配置 RIPng

```
R1(config)#ipv6 router rip 1
R1(config-rtr)#interface GigabitEthernet0/0/0
R1(config-if)#ipv6 rip 1 enable
R1(config-if)#interface GigabitEthernet0/0/2
R1(config-if)#ipv6 rip 1 enable
```

（2）在路由器 R2 上配置 RIPng

```
R2(config)#ipv6 router rip 1
R2(config-rtr)#interface GigabitEthernet0/0/0
R2(config-if)#ipv6 rip 1 enable
R2(config-if)#interface GigabitEthernet0/0/2
R2(config-if)#ipv6 rip 1 enable
```

（3）在路由器上查看 RIPng 路由

查看路由器 R1 的路由表：

```
R1>show ipv6 route
IPv6 Routing Table - 6 entries
<output omitted>
R      2021:806:718:B::/64 [120/2]
         via FE80::2, GigabitEthernet0/0/2    //路由器 R1 通过邻居路由器 R2 学习到 1 条 RIPng 路由
<output omitted>
```

查看路由器 R2 的路由表：

```
R2>show ipv6 route
IPv6 Routing Table - 10 entries
<output omitted>
R      2021:806:718:A::/64 [120/2]
         via FE80::1, GigabitEthernet0/0/2    //路由器 R2 通过邻居路由器 R1 学习到 1 条 RIPng 路由
<output omitted>
```

以上输出表明，路由器 R1 和 R2 通过 RIPng 学习到了对方的路由。

步骤三：配置 OSPFv3

（1）在路由器 R2 上配置 OSPFv3

```
R2(config)#ipv6 router ospf 1
R2(config-rtr)#router-id 2.2.2.2
R2(config-rtr)#interface GigabitEthernet0/0/1
R2(config-if)#ipv6 ospf 1 area 0
R2(config-if)#interface Serial0/2/0
R2(config-if)#ipv6 ospf 1 area 0
```

（2）在路由器 R3 上配置 OSPFv3

```
R3(config)#ipv6 router ospf 1
R3(config-rtr)#router-id 3.3.3.3
R3(config-rtr)#interface GigabitEthernet0/0/0
R3(config-if)#ipv6 ospf 1 area 0
R3(config-if)#interface Serial0/2/0
R3(config-if)#ipv6 ospf 1 area 0
```

（3）在路由器上查看 OSPFv3 路由

查看路由器 R2 的路由表：

```
R2>show ipv6 route ospf
IPv6 Routing Table - 11 entries
```

```
        <output omitted>
   O    2021:806:718:D::/64 [110/65]
          via FE80::3, Serial0/2/0    //路由器 R2 通过邻居路由器 R3 学习到 1 条 OSPFv3 路由
```

查看路由器 R3 的路由表：

```
R3>show ipv6 route ospf
IPv6 Routing Table - 8 entries
<output omitted>
   O    2021:806:718:C::/64 [110/65]
          via FE80::2, Serial0/2/0    //路由器 R3 通过邻居路由器 R2 学习到 1 条 OSPv3F 路由
```

以上输出表明，路由器 R2 和 R3 通过 OSPFv3 学习到了对方的路由。

步骤四：配置 IPv6 静态路由

（1）在路由器 R3 上配置 IPv6 静态路由

```
R3(config)#ipv6 route ::/0 Serial 0/2/1
```

（2）在路由器 ISP 上配置 IPv6 静态路由

```
ISP(config)#ipv6 route 2021:806:718:A::/63 Serial 0/2/1
ISP(config)#ipv6 route 2021:806:718:C::/63 Serial 0/2/1
ISP(config)#ipv6 route 2198:1118:12::/126 Serial 0/2/1
ISP(config)#ipv6 route 2198:1118:23::/126 Serial 0/2/1
```

（3）在路由器上查看 IPv6 静态路由

查看路由器 R3 的路由表：

```
R3>show ipv6 route
IPv6 Routing Table - 9 entries
<output omitted>
   S    ::/0 [1/0]
          via Serial0/2/1, directly connected    //路由器 R3 添加了 1 条 IPv6 静态默认路由
<output omitted>
```

查看路由器 ISP 的路由表：

```
ISP>show ipv6 route
IPv6 Routing Table - 8 entries
<output omitted>
   S    2021:806:718:A::/63 [1/0]
          via Serial0/2/1, directly connected
```

```
S    2021:806:718:C::/63 [1/0]
        via Serial0/2/1, directly connected
S    2198:1118:12::/126 [1/0]
        via Serial0/2/1, directly connected
S    2198:1118:23::/126 [1/0]
        via Serial0/2/1, directly connected
<output omitted>                        //ISP 添加了 4 条 IPv6 静态直连路由
```

（4）在路由器 R3 上测试与 ISP 的连通性

```
R3>ping 2001:1026::521
Type escape sequence to abort.
Sending 5, 100-byte ICMP Echos to 2001:1026::521, timeout is 2 seconds:
!!!!!                        //路由器 R3 与 ISP 通信成功
Success rate is 100 percent (5/5), round-trip min/avg/max = 16/17/21 ms
```

步骤五：传播 IPv6 默认路由

（1）在路由器 R3 上传播 IPv6 默认路由

```
R3(config)#ipv6 router ospf 1
R3(config-rtr)#default-information originate
```

（2）在路由器 R2 上查看 IPv6 默认路由

```
R2>show ipv6 route ospf
IPv6 Routing Table - 12 entries
<output omitted>
OE2 ::/0 [110/1], tag 1
        via FE80::3, Serial0/2/0    //从路由器 R3 传播来的默认路由，被添加到路由器 R2 的路由表中
O    2021:806:718:D::/64 [110/65]
        via FE80::3, Serial0/2/0
```

（3）在路由器上测试与 ISP 的连通性

```
R2> ping 2001:1026::521
Type escape sequence to abort.
Sending 5, 100-byte ICMP Echos to 2001:1026::521, timeout is 2 seconds:
!!!!!                        //路由器 R2 与 ISP 通信成功
Success rate is 100 percent (5/5), round-trip min/avg/max = 16/24/27ms
```

进一步测试路由器 R1 与 ISP 的连通性。

```
R1> ping 2001:1026::521
Type escape sequence to abort.
Sending 5, 100-byte ICMP Echos to 2001:1026::521, timeout is 2 seconds:
......                          //路由器 R1 与 ISP 通信失败
Success rate is 0 percent (0/5)
```

以上输出结果表明，路由器 R2 可以访问 ISP，而路由器 R1 无法访问。因为路由器 R3 将默认路由传播给路由器 R2，而路由器 R2 没有将默认路由传给路由器 R1。

步骤六：重分布 IPv6 路由

（1）在路由器 R2 上重分布路由至 RIP 域

```
R2(config)#ipv6 router rip 1
R2(config-rtr)#redistribute ospf 1 metric 7    //RIP 度量 16 跳不可达，此处设置度量为 7
R2(config-rtr)#redistribute connected metric 7
```

（2）在路由器 R2 上重分布路由至 OSPF 域

```
R2(config)#ipv6 router ospf 1
R2(config-rtr)#redistribute rip 1
R2(config-rtr)#redistribute connected
```

（3）在路由器上查看重分布来的 IPv6 路由

查看路由器 R1 的路由表：

```
R1>show ipv6 route
IPv6 Routing Table - 10 entries
<output omitted>
    R    ::/0 [120/7]
             via FE80::2, GigabitEthernet0/0/2
    R    2021:806:718:B::/64 [120/2]
             via FE80::2, GigabitEthernet0/0/2
    R    2021:806:718:C::/64 [120/7]
             via FE80::2, GigabitEthernet0/0/2
    R    2021:806:718:D::/64 [120/7]
             via FE80::2, GigabitEthernet0/0/2
    R    2198:1118:23::/126 [120/7]
             via FE80::2, GigabitEthernet0/0/2    //路由器 R1 通过路由器 R2 学习到 4 条度量为 7 的 RIP 路由
<output omitted>
```

查看路由器 R3 的路由表：

```
R3>show ipv6 route ospf
IPv6 Routing Table - 12 entries
<output omitted>
OE2 2021:806:718:A::/64 [110/20], tag 1
     via FE80::2, Serial0/2/0
OE2 2021:806:718:B::/64 [110/20], tag 1
     via FE80::2, Serial0/2/0
O   2021:806:718:C::/64 [110/65]
     via FE80::2, Serial0/2/0
OE2 2198:1118:12::/126 [110/20], tag 1
     via FE80::2, Serial0/2/0       //路由器 R3 通过 R2 学习到 3 条度量为 20 的 OSPF 外部路由
```

步骤七：测试网络连通性

至此，实现全网互通，请读者自行完成测试。以下输出表明，主机 PC1 发出的数据包顺利到达服务器 Server2。

```
C:\>tracert 2021:806:718:D::254
Tracing route to 2021:806:718:D::254 over a maximum of 30 hops:
  1   0ms    0ms    0ms    2021:806:718:A::       //路由器 R1
  2   0ms    0ms    0ms    2198:1118:12::2        //路由器 R2
  3   9ms    9ms    4ms    2198:1118:23::3        //路由器 R3
  4   1ms    0ms    0ms    2021:806:718:D::254    //服务器 Server2
```

以下输出表明，从路由器 ISP 发出的数据包顺利到达服务器 Server1。

```
ISP>traceroute 2021:806:718:A::254
Tracing the route to 2021:806:718:A::254
  1   2021:719::              2msec    7msec    0msec    //路由器 R3
  2   2198:1118:23::2         9msec    9msec    17msec   //路由器 R2
  3   2198:1118:12::1         14msec   1msec    10msec   //路由器 R1
  4   2021:806:718:A::254     9msec    1msec    1msec    //服务器 Server1
```

至此，IPv6 综合实验案例顺利结束。本实验包括 IPv6 静态路由、动态路由协议 RIPng 和 OSPFv3，以及默认路由传播和路由重分布，请读者在实验过程中注意观察路由表的变化，体会每一步的功能以及对路由表产生的影响。

8.5 挑战练习

学习完本章的内容后，有细心的读者可能已经发现，我们在本章中提到的一些语法知识，

在后续示例配置中并没有应用。这其实是本书一大特色,保留一些非必须配置语法,让读者自行去发现并应用,从而进一步感受网络世界的魅力。接下来,让我们将通过 2 个趣味挑战练习,引导读者探索新的发现。

8.5.1 挑战练习一

挑战要求:挑战练习一实验拓扑如图 8-6 所示,在该网络拓扑中,4 台路由器通过串行接口实现互连。PC1 访问服务器 Server1 有 2 条路径可以选择,为确保采用更安全路径 R1-R2-R3,我们需要人为干预选路,现在就让我们一起来挑战吧。

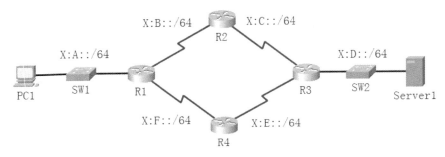

图 8-6 挑战练习一实验拓扑

任务要求:
- 请按图 8-6 所示搭建网络拓扑,要求路由器型号采用 4000 系列;
- 按图 8-6 为设备命名并配置 IPv6 地址,图中 X 代表 2198:1118:2188;
- 请分别采用 IPv6 静态路由和动态 OSPFv3 两种方法实现全网互通;
- 实现全网互通后,请对路由器选路进行人工干预,确保 PC1 与 Server1 间通信优先选择路径 R1-R2-R3,当路由器 R2 出现故障时,将切换至路径 R1-R4-R3,确保来回路径一致;
- 在主机 PC1 与服务器 Server1 上采用 **tracert** 命令验证所选路径;
- 保存路由器的配置文件,并将其上传配置到远程服务器进行备份。

请读者开动脑筋,根据任务要求独立完成本实验。

8.5.2 挑战练习二

挑战要求:挑战练习二网络拓扑如图 8-7 所示,由该图可知,网络采用 5 台三层交换机实现网络互联,3 台服务器分别提供 WEB、DNS 和 TFTP 服务。现在就让我们根据任务要求一起来完成这个挑战实验吧。

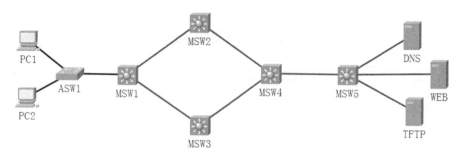

图 8-7 挑战练习二拓扑

任务要求：
- 按图 8-7 所示搭建拓扑，要求三层交换机型号均采用 3650 系列；
- 图 8-7 中三层交换机间以及其与服务器间均使用不同网段，采用光纤实现千兆位互连；
- 按图 8-7 所示，请采用 2198:1118:930::/48 网络前缀为全网规划和配置 IPv6 地址；
- 服务器地址为 X::254，服务器网关为本网最大地址，X 代表所属网络前缀；配置服务器使其提供相应的网络服务，WEB 服务器域名为 www.xllffy.com；
- 配置多区域 OSPFv3，实现全网互通，其中 MSW1～MSW4 属于 Area 0，MSW4 和 MSW5 属于 Area 1，MSWx 的路由器 ID 采用 x.x.x.x（x=1,2,3,4,5）；
- 配置三层交换机最多支持 2 个用户同时采用 SSH 方式登录，用户名为 xllffy，口令为 ytvc1716，特权口令为 BRACHN808，要求采用密文口令；
- 配置 IPv6 ACL，MSW5 禁止除所连服务器之外的所有用户远程登录；服务器禁止任何 ping 包，允许其他数据包通过；
- 保存三层交换机的配置，并将其上传配置到 TFTP 服务器进行远程备份。

请读者根据任务要求，独立完成本实验。

8.6 本章小结

本章内容到此结束。本章主要内容包括 IPv6 静态路由、IPv6 动态路由、IPv6 ACL 以及 IPv6 综合案例。本章通过 4 个场景来帮助读者掌握 IPv6 静态路由和动态路由协议的配置方法及应用。通过本章内容的学习，我们发现 IPv6 路由与 IPv4 路由间有许多相似之处，本章中给出的 4 个应用场景、1 个综合案例和 2 个挑战练习，可以帮助读者熟练应用 IPv6 技术规划和部署企业网络。

反侵权盗版声明

电子工业出版社依法对本作品享有专有出版权。任何未经权利人书面许可，复制、销售或通过信息网络传播本作品的行为；歪曲、篡改、剽窃本作品的行为，均违反《中华人民共和国著作权法》，其行为人应承担相应的民事责任和行政责任，构成犯罪的，将被依法追究刑事责任。

为了维护市场秩序，保护权利人的合法权益，本社将依法查处和打击侵权盗版的单位和个人。欢迎社会各界人士积极举报侵权盗版行为，本社将奖励举报有功人员，并保证举报人的信息不被泄露。

举报电话：（010）88254396；（010）88258888
传　　真：（010）88254397
E-mail：dbqq@phei.com.cn
通信地址：北京市海淀区万寿路 173 信箱
　　　　　电子工业出版社总编办公室
邮　　编：100036